Merry Christmas

Love Paul and Elizabeth

25/12/80

ORIGINS

ORIGINS

A
Skeptic's Guide to
the Creation of Life on Earth

Robert Shapiro

HEINEMANN : LONDON

William Heinemann Ltd
10 Upper Grosvenor Street, London W1X 9PA

LONDON MELBOURNE TORONTO
JOHANNESBURG AUCKLAND

First published in Great Britain 1986
Copyright © 1986 by Robert Shapiro
ISBN   0 434 69520 3

Printed and bound in Great Britain by
Billing & Sons Ltd, Worcester

Contents

Preface

Library shelves groan with the accumulated weight of books on the origin of life. This topic undoubtedly was discussed even before the invention of writing. Since that event, authors have not been timid in providing answers to the riddle of our origins. Why then am I adding yet another book to this collection? Because a need exists for a clear explanation, comprehensible to the general public, of what science does and does not understand about how life first began.

Remarkably, no adequate scientific explanation to the problem has emerged, despite the wealth of suggestions and the rapid progress made by science in many other areas. Not only the specific details concerning the start of life on earth are missing; at least one fundamental principle has not yet been grasped. Many books on the subject, in their eagerness to advocate a particular favored solution, fail to educate the reader on this cardinal point.

Despite the absence of a complete answer, a fascinating tale remains to be told. In what ways have the proposed theories proved inadequate? What motives have moved the proponents of various solutions to declare that their answers are the final ones? Many quarrels have arisen between advocates of competing positions, from a debate between scientist-priests in the eighteenth century, through episodes in the cold war, up to the continuing Creationist controversy of today. The history of these very human encounters provides an important background for understanding both the social state of the field today and the scientific quandary that it faces. For ultimately, from the study of the existing alternatives and the arguments used to advance them, the central problem emerges.

Once this problem has been stated and confronted, the way is left open for new speculations. I have not resisted the temptation of producing my own suggestions. They are presented not as final truths but rather as a stimulus to further investigations, to avoid encumbering the field with additional dogma. Finally, suggestions are made for experiments which may help resolve the outstanding difficulties.

The book opens with a Prologue intended to give the reader some notion of the immense diversity of the ideas that have been proposed for the origin of life. They include the well-publicized one that involves lightning and a soup of chemicals on the early earth, and others that invoke ancestors made of clay, the arrival of life from outer space, and intervention by an intelligent creator (proposed within a scientific rather than religious framework).

For a deeper examination of these and other possibilities, some background is needed. Chapter 1 distinguishes mythological and scientific approaches to the problem, pointing out the important criteria that a satisfactory scientific answer must meet. In Chapter 2, the principal features of life at the levels of cells and molecules are described, using a device that allows these features to be visualized. Chapter 3 considers the earlier history of life on this planet, as deduced from the fossil record and radioactive dating methods. Those readers who are familiar with one or more of these subjects may choose to move past them to the material further on.

In Chapters 4 to 10, theories prominent today are considered, both in their content and their often turbulent background. They are compared to the standard that we established in Chapter 1 for the good practice of science. We note what they tell us and fail to tell us about the origin of life. In Chapter 11, their method of coexistence today is illustrated, through an account of a prominent international conference. The remainder of the book provides speculative suggestions concerning the origin and development of life, and suggests studies that may lead us to the answer.

The guiding spirit on this entire quest is the scientific approach, and the way in which it views and explores the world. If the reader takes from this book not only a sense of wonder at the unsolved riddle of our existence, but also a preference for doubt in place of dogma and a keen appreciation for the proper practice of science, then I will have achieved my purpose.

I am indebted to those who helped me in the preparation of this book. A number of fellow scientists were willing to discuss their ideas with me, often at length. They included Graham Cairns-Smith, Francis Crick, Donald DeVincenzi, Gerald Feinberg, Jim Ferris, Sidney Fox, Hyman Hartman, Clifford Matthews, Stanley Miller, Leslie Orgel, Cyril Ponnamperuma, Bill Schopf, Alan Schwartz, Charles Thaxton, and David Usher. I wish to thank them for their time and attention.

I am also grateful to my agents, John Brockman and Katinka Matson, for their encouragement along the way, and to my typists, Meredith Storer and Pat Smith. Finally, I wish to thank my editor, Arthur Samuelson, for his valuable suggestions about the overall shape of the book.

Prologue: In the Beginning

When I was a child I wondered where I had come from, so I asked my mother. She replied that I had grown in her stomach. Later I found that this was a fair approximation of the truth (we won't quibble about the difference between a stomach and a uterus), but at the time I couldn't accept it. After all, she frequently told me that if I ate hamburgers as often as I wanted, day after day, they would grow in *my* stomach. I disliked the thought that my beginning was that of a hamburger that had gotten out of control, so I put aside the theory of the internal origin of babies and turned to other sources.

I grew up in New York City in the 1940s, at a time when public sensibilities did not permit a rich supply of information about sex and birth into the media. Comic strips and movie cartoons showed storks delivering babies, an external origin theory. This seemed unlikely. The skies were not filled with winged messengers, despite the numerous baby carriages in my neighborhood. Hardly any birds at all, except pigeons and dusty English sparrows, were to be seen, and they were too small to carry the load.

Even if storks did deliver babies, the problem of where they got the infants would remain.

I had no chance for direct observation, as there were no pregnancies among close relatives at the time, so I invented fanciful tales of my own. My mother had sometimes spoken of God, though we observed no formal religion. I imagined that when conditions were right, which was determined by some planning board in heaven, babies simply appeared miraculously, next to their mother. Perhaps the mother had some sort of warning in advance, enough to allow her to get to the hospital, where the baby could receive proper care on arrival.

I had to abandon this idea when I learned the actual facts from peers on the street. The news must have swept into our neighborhood like an invading army, for I heard the blunt graphic details from so many directions at once that I was overwhelmed. Much later, as an adult, I had the chance to dispel the last elements of mystery about the process, as I witnessed at first hand the birth of my own son Michael. The solution of this problem, however, only raised a much larger one. I wondered instead about the origin of my most distant human ancestors and the beings who may have come before them, back to the start of life itself.

As I chased this new problem, I found myself once more in the position of a child: I had no chance for direct observation of the process. Answers existed in overabundant supply, but none was convincing on its own authority. I could choose to believe that life grew like some hamburger within the belly of Mother Earth, or that it had been brought by winged messengers from outside. My own childhood theory of miraculous arrival had of course been thought of by others much earlier, and now had enormous institutional sanction. Most religions taught that life first began in this manner.

I did not wish a religious answer, so I turned to my peers again. I no longer found them on the street; I now identified with those who worked in scientific laboratories. This time, however, I could not be helped by them. It was too early; the news had not yet reached the neighborhood. They had not yet agreed, at least on the most crucial point of the story.

Science does present us with a coherent account of the development of life on this planet. If I accept that I grew from a fertilized egg into a baby and ultimately into the adult human that I am, then I can believe that a single cell on the early earth could evolve to give the life we see today. It is very surprising to think that a single cell could contain enough information to make *me*. Once I get past this hurdle, then I am ready for the larger idea that some primitive bacterium had the potential, over four billion years, to become all of us.

Scientific explanations flounder, however, and possibilities multiply when we ask how this first cell arose on earth. Competing theories abound—which seems always the case when we know very little about a subject. Some theories, of course, come labeled as The Answer. As such they are more properly classified as mythology or religion than as science.

Having said this much, I cannot claim that this book will present The Answer. I write it for today; tomorrow may be different. The question of our origins is a glorious one, however, which has excited mankind throughout history, and which presents a story well worth telling. By a study of the answers that have been given, we can, with care, extract a partial answer to the origin of life for ourselves.

We cannot proceed without a point of view. Because of something in my upbringing, I have found that I tend to doubt most new information when I hear it. I am different in this way from many others who are more trusting. But there are times when I hear a piece of news that I like particularly, or have guessed at in advance, and I embrace it wholeheartedly, without doubt, in contrast to my usual style.

I have invented a protagonist who will accompany us on our quest. He embodies the skeptical tendencies that I find in myself, but does so consistently, without human flaws. He will be called the Skeptic. He will be summoned from time to time throughout the book when a particularly choice balloon requires deflation, but in this chapter he will be given a larger role. Before he starts, however, he will need a discipline, a set of rules by which to apply his skeptical outlook. His selection will be the one I made for myself in adolescence, namely science.

We will start with a shaggy dog joke. My childhood friends and I spent hours on street corners and playground benches in a competition to invent the longest and most anticlimactic jokes possible. The jokes usually had a standard format, which we padded with whatever adventures we could devise at the moment. One format involved the search for the world's most shaggy dog, hence the name I chose above. Another, more relevant here, was the quest for the answer to an ultimate question, such as the meaning of life.

In a typical tale of this sort, the hero would seek a supreme authority who knew the answer. He would be told, for example, to locate a wise Guru who lived atop an inaccessible mountain in the Himalayas. Hours of real time would pass as we heard of the years-long adventures of the hero in trying to locate the Guru. Finally, he would locate this person and find that his appearance satisfied all expectations. The Guru would have long robes, a benign wrinkled face, and an immense gray beard. He would sit in a position of perpetual meditation.

The hero, of course, would immediately put the question to him and receive an answer like "Life is a fountain." This threw our hero into confusion and anxiety, and after some struggle with himself he would blurt out, "That can't be the answer, that life is a fountain." The Guru would then calmly reply, "So, life is not a fountain."

Our hero in this book, of course, is the Skeptic, and his question, the origin of life. We will again skip the adventures and move to the point of confrontation. The Guru, in the meantime, has meditated upon his previous encounters and learned from them. Longer answers were expected by his questioners. Further, they often did not like the answer they received. He decided that he would work with them at greater length, to help them find an acceptable solution.

When the Skeptic asked his question, the Guru made the following offer: The Guru would try to provide an answer. If the Skeptic did not like what he heard, he was welcome to return on the following day, and the Guru would try again. The Guru was willing to keep this up for a week, if necessary. Then he would have to attend to other matters.

The offer was accepted, and the Guru began his first answer on that same day, which was a Monday.

Monday's Tale

The first living being that we know of was called Father Raven. He created all life on earth, and was the origin of everything. He was a holy life-power who began in the shape of a human and later became a raven.

He awoke suddenly to consciousness and found himself crouching in the darkness. He did not know how he had come into being, or where he was. Everything around him was dark, and he could not see anything. He groped about in the darkness, but felt only dead clay. He then explored his face and body with his hands, and found that he was a human being, a man. In addition, a hard little knot existed above his forehead, which would one day grow into a beak, but he did not know this.

Father Raven crept over the clay to explore his surroundings. As he moved, he encountered a hard object, which he buried on impulse. Continuing on his trip, he abruptly came to an edge and turned back. Suddenly he heard a whirring noise and felt a little creature settle on his hand. He explored it with his other hand and recognized it as a sparrow. This little sparrow had been there first, and had approached him in the darkness. He had not been aware of it until it touched him.

Father Raven now resumed his search and returned to the place where he had buried the object. It had grown roots and become a bush. Other bushes and grass now grew nearby in the naked clay. The man felt lonely, so he formed a figure from the clay, which resembled his own, and waited. The new human being came alive and started to dig about restlessly. It had a hot temper and a violent attitude. Father Raven did not like it, so he dragged it to the edge and threw it into the abyss. This being later became the evil spirit, the source of all evil on earth.

Father Raven crept back to the growing bushes and found that they had become a forest of trees. He crawled about to explore his dark land, but encountered water in all directions except the one that led to the abyss. The little sparrow had been flying above his head all the time, so he asked it to go down and find out what was there. The sparrow did so, and on returning, reported that a new land was there, which had just crusted over.

They had been in the land called heaven; the slightly younger land below, Father Raven named the earth. He examined the sparrow and felt the construction of its wings. He made a similar pair for himself, from twigs in the forest, and placed them on his shoulders. The twigs were transformed into real wings, while he himself grew feathers and a beak. He had become a big black bird, and called himself Raven.

Father Raven and the sparrow made the long flight from heaven to earth, which exhausted them. When he had recovered, Father Raven planted the new land, as he had done in heaven, and then created human beings. Some say that he made them out of clay, as he had made the first being in heaven. Others claim that he created man by chance, which would be stranger than if he had done so deliberately. He had planted some pods, then later opened one and the first human being popped out. Father Raven then created all other beings.

When he had populated the earth, the Raven assembled the humans and said to them, "I am your Father. You owe your land and your very existence to me. Do not forget me." Then he returned to heaven.

It had been dark all this time. With some firestones he now created the stars, and a great fire which lit the earth. In this way the earth, humans, and all life came into existence, but before them came the Raven, and the little sparrow came even before him.

When the Guru had finished his tale, the Skeptic asked about the source of this particular story. He learned that it was told by an Eskimo, Apakag, to a Scandinavian explorer, Knud Rasmussen, on the shores of the Arctic Ocean. It was said to represent the ancient wisdom of his people.

The Skeptic then asked what reasons there might be to believe that this story gave an accurate account of events concerning the origin of life. The Guru replied that we had only the word of the Eskimo. Many cultures, he added, have devised creation myths.

The Eskimo version had some unusual features, such as the state of confusion of the creating power, that made it interesting to him. Creation myths varied greatly in their details, but each asserted its own validity.

How, then, the Skeptic inquired, could a choice be made among them? He was told that it was only a matter of personal preference.

He then declared that none of them would do. He was not interested in mythology. Rather, he wished to have an answer from a discipline in which different points of view were compared and the correct one selected by common agreement. Science had these characteristics. Could the Guru present him with an account of the origin of life on which scientists had agreed? The Guru said that he would do that on the next day, Tuesday.

Tuesday's Tale

Nature is unified, the Guru began. It represents a vast, infinite entity, which in itself may be considered alive. Life and death are simply different aspects of this same unity. As living beings may readily become not alive, so may nonliving matter transform itself and give rise to living beings. The simpler creatures, in particular, can be formed with great ease.

No profound proof is needed for this truth. It has been observed by wise and practical men since ancient times. Aristotle and his followers noted that fireflies emerged from morning dew, and many types of small animals arose from the mud at the bottom of streams and ponds. The process later came to be called spontaneous generation. The philosopher René Descartes analyzed it and suggested that it was the result of heat agitating the subtle and dense particles of putrefying matter.

Many other noted scientists and philosophers have endorsed spontaneous generation, including Thomas Aquinas, Francis Bacon, Galileo, and Copernicus. A great list of organisms has been compiled that may be made in this manner. John Needham demonstrated that microorganisms arise spontaneously in the most carefully sterilized broths. Others have observed that worms arise in timber, beetles in dung, and mice from river mud.

The Nile, in particular, has very fertile qualities. Great literature has borne witness to this. In Shakespeare's *Antony and Cleopatra*, Lepidus tells Antony: "Your serpent of Egypt is born of the mud, by the action of the Sun, and so is your crocodile."

More exotic recipes can work to the same end. Jan Baptist van Helmont, a seventeenth-century Flemish biologist, developed a procedure for the production of mice from a mixture of wheat and sweaty underwear. The mice appeared in adult form, and could cross-breed with normal mice.

Scientists have agreed for many centuries that there is no problem to the origin of life. All manner of creatures arise continually, everywhere around us.

The Guru had completed his presentation, but the Skeptic looked puzzled, as if he had expected more. He finally commented that he thought spontaneous generation had been abandoned. The Guru agreed that it was disbelieved at the present time. However, it had been accepted, almost without dissent, for many centuries. And the Skeptic had, after all, requested an account on which men of science had agreed.

The Skeptic protested that he was not concerned with the theories of the past. He wanted an answer on which there was full agreement today. When he was told that no such consensus existed, he then requested that explanation which had received the most acceptance. The Guru promised to tell it on Wednesday, and began promptly the next morning.

Wednesday's Tale

The earth was four billion years old. The sky looked much as it does now, but its gases were strange. In place of oxygen, the atmosphere contained methane, hydrogen, and fumes of ammonia.

Life was absent. The planet was covered with a shallow, sterile sea. Bleak islands provided the only lands; no continents existed yet. The landscape was not quiet, however. Roaring volcanoes sent forth lava. Steam and poisonous gases escaped into the air from hot springs bubbling nearby.

Now and then a thunderstorm lashed out at our planet. Flashes of lightning illuminated the scene. The electrical discharges agitated the gases of the atmosphere, causing them to combine with each other and with water. Strange new molecules were formed, called amino acids and nucleotides. They had not been seen previously on the earth. They were the building blocks of living matter.

Gradually, more and more amino acids and nucleotides filled up the seas, creating a rich organic soup, more concentrated than a chicken broth. The molecules collided in the broth, and every so often stuck together. Larger and larger molecules were formed. In the course of hundreds of millions of years, all manner of molecules were created by random collision. Some had spiral shapes, other were spherical, and still others had long strands.

Finally, after billions of chance events, a molecule was formed that had the magical talent of copying itself. This magic molecule had two long chains of nucleotides twisted around one another. When the chains separated, each attracted nucleotides to itself and constructed a copy of its earlier partner. Two giant molecules then existed in place of one. Reproduction had taken place.

This replication process occurred over and over again. Soon offspring of the original parent molecule dominated the waters of the young earth. They were the earliest forms of life.

In the billions of years that followed, these early self-reproducing molecules evolved, and ultimately produced the variety of creatures that fill the earth today—germs, plants, mice, and men. Each creature is made of cells, and the cells are made of the same building blocks, amino acids and nucleotides. At the center of every living cell lies a descendant of the first living molecule. We now call it DNA.

This time the Skeptic looked almost satisfied at the end of the tale. He had encountered that story many times, in slightly different forms, in schools, museums, and the popular media. He liked this particular version, and was glad to hear that it was accepted by many scientists. What about the remainder? Would they soon come around as well?

The Guru agreed that this story had been told many times. He had taken his version from an account given by the astronomer Robert Jastrow in his book *Until the Sun Dies.* It was not likely, however, that the scientists who rejected this theory now would accept it in the future. In fact, there were more dissenters now than there had been twenty years ago.

The Skeptic asked why this was so. He was told that a growing number of scientists now believed that neither the atmosphere described nor the soup had ever existed. Laboratory efforts had also been made to prepare the magic molecule from a simulation of the soup, and thus far had failed.

Could life have started on earth without this atmosphere, the soup, and DNA? The Guru replied that a new idea had arisen which dispensed with these ingredients. He would present this story on the next day.

Thursday's Tale

Four billion years ago, the earth had rocks, water, and air, as it does now. Strange gases did not fill the air, nor did nucleotides and amino acids swim in the sea. The atmosphere had nitrogen and carbon dioxide, familiar to us today. Only oxygen, which we now need for life, was lacking.

Storms occurred, and rains did fall. The rocks weathered, dissolved, and sedimented. New soils and minerals were formed. Among them were the clays, which crystallized in many varied patterns. The different forms grew, fragmented, were carried downstream, and grew again. As circumstances changed, some spread widely and others disappeared.

Let us follow the adventures of three such clay forms, which we will call Sloppy, Tough, and Lumpy. Each had come to dominate the region in which he had been first deposited. Sloppy had a loose, open consistency. He had many open channels through which water could pass. The flowing waters deposited minerals, making more of Sloppy. He grew quickly. Tough had a dense, closed form. He stuck very well to the neighboring rocks, but hardly any water passed through him. His growth was very slow. Lumpy had a blend of the qualities of the other two. He was coagulated, with the consistency of a badly made custard pudding.

He grew at a moderate rate. The three clays had lived their lives in rather constant, dry weather.

One day the weather changed, and heavy rains fell. Sloppy had too loose a hold on his rock. He was swept away, and was never heard of again. Tough hung on quite nicely and carried on his marginal business pretty much as usual. He did not grow or spread, however, and played little role in future events. Lumpy was the one who adapted best to the situation. He broke into pieces. Parts of him remained in place. The remainder—his children—were swept downstream. Many were able to attain a new foothold on another suitable rock. When the floods subsided and normal conditions returned, the baby Lumpies resumed growth.

This cycle was repeated. New generations of Lumpies were raised, and evolved into improved versions. One day, a novel Lumpy appeared that had learned to build organic molecules—the type that we use in life today—into his structure. This practice spread and escalated. The clay-organic beings were better at survival than those made of clay alone, and further improvements could be made by reducing the clay content even more.

One day a distant descendant of Lumpy reached the logical end of this process. The last clay was discarded. He was no longer bound to the rocks and could float freely in the waters of the earth. Modern evolution had begun, from this first cell made only of organic chemicals.

The Skeptic frowned and asked whether they had not agreed to dispense with mythology. Yet here they were again, with beings made of clay. Only Father Raven was missing.

The Guru replied that, yes, the creation of life from clay was a feature of many mythical accounts, and had been a part of the spontaneous generation theory as well. However, the above tale was based on authentic science, developed by a chemist at the University of Glasgow, Scotland, named Graham Cairns-Smith. Much of it was speculation and few relevant experiments had been done, but it was science nonetheless.

The Skeptic then asked what support the theory had received from other scientists. He learned that its author was not yet well known in science. Only a small but growing number of followers favored this hypothesis.

If he had to hear speculation, the Skeptic said, he would prefer to entertain the ideas of the celebrated scientists of our time. The Guru agreed to honor his wish, on the next day.

Friday's Tale

A long time ago, in a star system far away, there lived a civilized race. Their sun was much like our own, but it predated ours by billions of years. The home planet of this race resembled earth in some ways—it had an atmosphere, oceans of water, and a pleasant climate. The planet was more massive than earth, however, and its large gravity permitted it to keep much of the cloud of hydrogen gas in which it had been born. This hydrogen atmosphere made it a very suitable place for the origin of life, unlike the earth.

Life began on this faraway world and evolved to more complex forms. Eventually, after billions of years, intelligent beings appeared. The process of evolution had started on this planet when the universe was young, and civilization existed there at the time when our own solar system and home planet had just been formed.

Here a sad note enters our story. These beings, whom we will call the Old Ones, learned that civilization could not endure on their planet. Eventually, their sun would become a red giant. Their own world would be engulfed and roasted. The Old Ones tried various strategies to save themselves. They explored nearby worlds, in their own solar system and neighboring ones, to find planets suitable for colonization, but none were found.

Probes were then sent by remote control to explore more distant stars. They reported back that a few worlds existed roughly similar to their own, but none had developed life. In a few cases, a soup of organic molecules had collected, but some factor or other was missing that was needed to complete the process.

The Old Ones then built starships. Expeditions lasting many generations would carry settlers to the new worlds. The closest ones were 100 light-years away. The best starships that they could devise would require 10,000 years for the trip, a number equal to a great many of their lifetimes. They were unable to develop methods of suspended animation which would work for that span of time. Instead, they sent small societies into space, hoping that the descendants of the pioneers would arrive to colonize the new planets. However, such societies proved unstable. The vessels returned, or were lost, after a few centuries.

At this point, the Old Ones recognized that they could not arrange the survival of their civilization. They settled for a lesser goal, the perpetuation of life itself. They did not expect that higher organisms could survive a 10,000-year trip through space, but bacteria could do this quite easily.

Special spaceships were built to house frozen bacteria on this long voyage. In each vessel many packets were stored, each containing billions of bacteria. Some of the microbes could use organic molecules as food, while others could thrive on minerals, and a number of species were able to make their own food, using light energy from a sun (photosynthesis). Each ship was targeted to a solar system known to contain a world suitable for life.

Four billion years ago, one such ship approached the earth. The target was recognized. The packets were released and their contents scattered over the surface of this planet. Many bacteria landed in unsuitable locations. A few found safe haven in a sea or pond. The most suitable species grew and evolved. We are their direct descendants, but the Old Ones were our godparents.

The Guru did not wait for a question at the conclusion of this tale, but added a historical note. Francis Crick had discussed this concept in great detail in his book *Life Itself*. Several other scientists had mentioned it briefly at earlier times. Crick was, of course, one of the most famous scientists of our time, a Nobel Prize winner. With James Watson, he was co-author of the Watson-Crick theory, the most important idea in modern genetics. Crick had made many other outstanding contributions to science.

These comments left the Skeptic a bit disconcerted. He thought for a while and asked a number of questions: Did Francis Crick really *believe* that life on earth began this way? How would we ever learn anything about the Old Ones? In any event, even if the story, which seemed science fiction, were true, it still had no solution to the ultimate question of the origin of life. How had the Old Ones come into being on their planet?

The Guru responded that Crick was not convinced that this series of events had taken place. He had only suggested it as an alternative to conventional theories. It was difficult to obtain evidence in its favor at this time. Crick considered the theory to be premature. And no, no suggestion was made concerning the origin of the first life in the galaxy.

Yet, the Guru continued, there was another theory, similar in some ways to Crick's. It also postulated an extraterrestrial origin for life here, four billion years ago. Another famous British scientist was the principal author. He had not received the Nobel Prize, but he had been knighted. He was the astronomer Sir Fred Hoyle.

Hoyle was convinced of the validity of his theory. He offered evidence in support of it, and carried origins back to an ultimate answer. The Guru would continue with it tomorrow, if the Skeptic cared to hear it. The Skeptic agreed.

Saturday's Tale

Life first arrived on earth from space, in the form of living matter, bacteria and viruses. Cells, viruses, and bits of genetic material have continued to arrive throughout the history of our planet, causing many of the biological advances attributed to Darwinian evolution.

The living material that reached us was ejected earlier from another solar system. It traveled here through interstellar space, riding on the pressure of starlight, until it encountered a vast cloud of gas. This cloud eventually collapsed, giving birth to our own solar system.

Our sun was very hot and bright in its early history, so the temperature at the distance of the earth's present orbit was like that of a blast furnace. Farther out, in the vicinity of the orbits of Uranus or Neptune, the temperature was more favorable, about

20° C (68° F), ideal for the processes of life. At this temperature, the bacteria multiplied rapidly, using chemicals in the cloud as food. Comets were forming at this time, and some of the bacteria found homes within them, reproducing until enormous numbers were present. Other bacteria escaped to interstellar space, starting a journey to another solar system.

The sun cooled down, and the planets were formed. Many comets remained out beyond the orbit of Neptune, where a great cold now existed. The bacteria in the comets froze, with their life processes entirely suspended, and have remained so for four billion years.

From time to time, however, a comet was deflected into a new orbit by the gravity of a passing star and entered the inner part of the solar system. As the comet approached the sun, frozen material at its surface thawed and evaporated. Cells and viruses were released into space with other particles and bombarded the earth. In addition to this infall, whole comets occasionally made soft landings on the earth and on other planets with atmospheres, such as Mars. During the early history of the earth, living cells from cometary sources were continuously arriving on the surface. Many perished, but several types survived and established themselves. Thus began life on earth.

This influx from comets continued through the ages. New biological material produced evolutionary advances, but not all of the effects were beneficial. In recent history epidemic diseases, including many outbreaks of influenza, were caused by infection from cometary sources.

This story has not yet told how the bacteria and viruses originated in space. Even such simple forms of life are far too complex to have arisen by random chemical reactions in a soup. They were designed by a higher intelligence, perhaps a being based on silicon chemistry. Even more intelligent beings stand behind the ones who created us. Such beings would be able to control the basic rules of physics themselves, and determine many features of the universe.

A chain of intelligent beings exists, leading up until we reach the ultimate intelligence, God, who is the universe itself. God equals the Universe.

* * *

A long silence followed the conclusion of this narrative, then the Skeptic asked the expected questions. He inquired about the extent of the evidence, and the nature and breadth of the support that these ideas had received from other scientists. He commented on the lack of detail concerning the chain of higher intelligence. Was this science or religion?

The Guru reported that Hoyle and his collaborator, Chandra Wickramasinghe, essentially stood alone in defense of their theory, despite having published an entire series of technical papers. God and higher intelligences were not dealt with in these detailed works, but rather in a popular book. Limited parts of their evidence were accepted by some conventional scientists, but most of it had drawn heavy criticism.

Another group existed, however, that considered themselves to be scientists. They strongly endorsed certain parts of this theory, in particular the rejection of the idea of a chemical origin of life in favor of the concept of an ultimate creator. The views of this group have immense popular support, the Guru continued. (Despite his Himalayan abode, the Guru found ways of informing himself of the most recent public events.) In a 1982 Gallup poll, 44 percent of the American public endorsed their position on the creation of man, and presumably of life. Hoyle and Wickramasinghe have in effect worked together with this group in certain causes. The Guru would present the view of this group on origins in his final tale, to begin on Sunday morning.

Sunday's Tale

In the beginning God created the heaven and the earth.

And the earth was without form, and void; the darkness was upon the face of the deep. And the Spirit of God moved upon the face of the waters.

And God said, Let there be light: and there was light.

And God saw the light, and that it was good: and God divided the light from the darkness.

And God called the light Day, and the darkness he called Night. And the evening and the morning were the first day.

And God said, Let there be a firmament in the midst of the waters, and let it divide the waters from the waters.

And God made the firmament, and divided the waters which were under the firmament from the waters which were above the firmament: and it was so.

And God called the firmament Heaven. And the evening and the morning were the second day.

And God said, Let the waters under the heaven be gathered together unto one place, and let the dry land appear: and it was so.

And God called the dry land Earth; and the gathering together of the waters called he Seas: and God saw that it was good.

And God said, Let the earth bring forth grass, the herb yielding seed, and the fruit tree yielding fruit after his kind, whose seed is in itself, upon the earth: and it was so.

And the earth brought forth grass, and herb yielding seed after his kind, and the tree yielding fruit, whose seed was in itself, after his kind: and God saw that it was good.

And the evening and the morning were the third day.

And God said, Let there be lights in the firmament of the heaven to divide the day from the night; and let them be for signs, and for seasons, and for days, and years:

And let them be for lights in the firmament of the heaven to give light upon the earth: and it was so.

And God made two great lights; the greater light to rule the day, and the lesser light to rule the night: he made the stars also.

And God set them in the firmament of the heaven to give light upon the earth,

And to rule over the day and over the night, and to divide the light from the darkness: and God saw that it was good.

And the evening and the morning were the fourth day.

And God said, Let the waters bring forth abundantly the moving creature that hath life, and fowl that may fly above the earth in the open firmament of heaven.

And God created great whales, and every living creature that moveth, which the waters brought forth abundantly, after their kind, and every winged fowl after his kind: and God saw that it was good.

And God blessed them, saying, Be fruitful, and multiply, and fill the waters in the seas, and let fowl multiply in the earth.

And the evening and the morning were the fifth day.

And God said, Let the earth bring forth the living creature after his kind, cattle, and creeping thing, and beast of the earth after his kind: and it was so.

And God made the beast of the earth after his kind, and cattle after their kind, and every thing that creepeth upon the earth after his kind: and God saw that it was good.

And God said, Let us make man in our image, after our likeness: and let them have dominion over the fish of the sea, and over the fowl of the air, and over the cattle, and over all the earth, and over every creeping thing that creepeth upon the earth.

So God created man in his own image, in the image of God created he him; male and female created he them.

And God blessed them, and God said unto them, Be fruitful, and multiply, and replenish the earth, and subdue it: and have dominion over the fish of the sea, and over the fowl of the air, and over every living thing that moveth upon the earth.

And God said, Behold, I have given you every herb bearing seed, which is upon the face of all the earth, and every tree, in which is the fruit of a tree yielding seed; to you it shall be for meat.

And to every beast of the earth, and to every fowl of the air, and to every thing that creepeth upon the earth, wherein there is life, I have given every green herb for meat: and it was so.

And God saw every thing that he had made, and, behold, it was very good. And the evening and the morning were the sixth day.

Thus the heavens and the earth were finished, and all the host of them.

And on the seventh day God ended his work which he had made; and he rested on the seventh day from all his work which he had made.

And God blessed the seventh day, and sanctified it: because that in it he had rested from all his work which God created and made.

*   *   *

"I've heard that one before," the Skeptic said, "but that *certainly* is religion. It may be very good religion. But that's not what I'm looking for. I came all this way for a scientific answer, not religion or a myth. I thought I explained that to you. This one doesn't count. I want a different story."

The Guru was unmoved by this plea. Yes, many people accepted this account as religion. But the group he had mentioned, the Creationists, maintained that it was science, and insisted that it should be taught in science classes in American schools. He had presented this material not as religion, but to represent their viewpoint that it was science.

In any event, he had no further time to spare. He suggested to the Skeptic that if he wished a scientific answer to his question, then he should master the material on his own, rather than appeal to authority, even so wise an authority as a Guru. But first, he added, it might be prudent to learn more about the nature of science, and the distinction between it and religion and mythology.

We shall follow the Guru's advice.

one

Doubt and Certainty

Living creatures differ so amazingly from the inanimate world around them that we cannot help wondering how life began, and came to its present forms. Was the start of life an accident, or the inevitable outcome of natural laws, or perhaps the deliberate act of a powerful supernatural being? The answer to this question matters deeply to us, as it affects not only how we view the meaning of our own lives but the larger purpose of life itself.

The question of the origin of life has therefore been asked as long as humans have existed, and each society has provided an answer for it. The usual form of these answers has been that of a myth: an account that declares its own validity, rather than attempting to demonstrate it by some objective procedure. Such myths commonly have been incorporated within a larger religious framework that provides guidance about many aspects of human existence.

In recent times, an alternative way of dealing with reality has captured the imagination of humanity: science. The development of the modern scientific view of the universe has been a glorious intellectual enterprise of the human race. Many events that at one time seemed complex and obscure, from the movements of the stars to the basic operations of our own bodies, have been made comprehensible to us. Further, this knowledge has been used to bring much of nature under our control on an everyday basis. Our ancestors waited patiently for the dawn, but we can turn on the lights with a movement of one finger. They suffered chronic maladies, but often we need only swallow a pill to make the ache go away.

These triumphs of technology testify to the power of the scientific approach. They lead us to expect that science can also tell us how life first began. Those scientists most concerned with origin-of-life research have in fact provided such an account for us. It tells of an early earth covered with roaring volcanoes, where thunder and lightning storms flash in an atmosphere of strange gases. Many chemicals are formed, which dissolve in the seas to create a mixture called the prebiotic soup. This pregnant broth contains most of the ingredients necessary for life. One day, by chance, it gives birth to a chemical with the marvelous power to reproduce itself. It does so, filling the broth with its descendants, and Darwinian evolution begins.

This picture has remained in place for a generation. We learn it in high school science classes and encounter it again in museums and the media. Popular articles and press releases inform us that yet another piece in the nearly completed picture has fallen into place. On closer inspection, however, we find that not all is well in the field. It does not rest securely, like our understanding of the movements of the planets or the circulation of the blood.

Advocates of the ruling theory disagree passionately on an important detail: the chemical identity of the first self-replicating molecule. The majority supports the nucleic acids, the carriers of heredity today. A vocal dissenting minority prefers the proteins, another important contemporary class of biochemicals. Most recently, a radical faction has suggested that clay minerals, which usually suggest pottery to us rather than reproduction, played this vital initial role.

Some eminent scientists have turned away from all of these efforts to describe the start of life on earth and proposed a dramatic alternative: it started elsewhere and came here. One of them, Sir Fred Hoyle, has further insisted that a higher intelligence, chemically unrelated to us, was the creator of our kind of life. In advancing this idea, he has made common cause with a much larger group who wish to invoke the biblical Creator for the same purpose, not as religion but in the guise of science.

In the course of this book we shall repeat, on a broader scale, the quest initiated by the Skeptic in the Prologue: we desire the best available scientific status report on the origin of life. We shall see that adherents of the best-known theory have not responded to increasing adverse evidence by questioning the validity of their beliefs, in the best scientific tradition; rather, they have chosen to hold it as a truth beyond question, thereby enshrining it as mythology. In response, many alternative explanations have introduced even greater elements of mythology, until finally, science has been abandoned entirely in substance, though retained in name.

To fulfill the purpose of our quest, we must return to the proper practice of science. In particular, we will affirm the value of doubt. This essential element is often overlooked when science is presented to the public. In everyday use, a statement that something is scientific means that it is correct, proven beyond question. Who dares argue with a scientific fact? The earth is round and moves around the sun. The universe is made of atoms, which combine into molecules. We need not trouble ourselves further with these matters. So authoritative is the name of science that the term is added to mundane processes to give them an air of precision, as in "upholstery science," or to validate dubious areas of inquiry, as in "psychic science." The scientific word is the final word.

In the origin-of-life field, we face a host of contending theories, each claiming to be the one valid scientific answer. In the course of this book, we shall hold them up to the rigorous standards of evidence used in the main body of contemporary science. We shall learn what has been understood about the history of life and what important problems remain unanswered, and even unexplored. We can then sketch out some possible solutions and suggest how the missing information may be acquired.

Before we attempt this feat, we must become more familiar with our tools. In the remainder of this chapter, we shall learn more about the best practice of science and the philosophy that motivates it.

Science: The Realm of Doubt

I have chosen this title to make the strongest possible contrast between the common view of science described above and its essence. Science is not a given set of answers but a system for obtaining answers. The method by which the search is conducted is more important than the nature of the solution. Questions need not be answered at all, or answers may be provided and then changed. It does not matter how often or how profoundly our view of the universe alters, as long as these changes take place in a way appropriate to science. For the practice of science, like the game of baseball, is covered by definite rules.

Neither science nor baseball can be done well unless the players agree to follow the rules, or at least do not vary them at their whim. In baseball, the runner goes from home plate to first base after he hits the ball. He may stretch the rules by running outside the baseline, and get away with it, but his direction is clear. If a player chose to run from home directly to third base, he would be called out. If he insisted that his own direction was correct, he would be thrown out of the game. In this book we will encounter arguments that are presented as science, but those who make them are running from home toward third base. They are seeking answers in their own way, but that way is not within science.

In the origin-of-life field, a particular theory or point of view is frequently elevated to the status of a myth. It is then treated only as a doctrine to be validated, and not one to be challenged. It is important that we recognize such cases, so we will pause to consider the proper use of myths and their contribution to human thought on the origin of life.

Mythology: The Realm of Certainty

My encyclopedia traces the term "myth" back to the ancient Greek *mythos*, which means "word," in the sense that it is the decisive, final word on the subject. A myth presents itself as an authoritative account of the facts, which is not to be questioned, however strange it may seem. The opposite side of this coin is *logos*, the Greek term for an account whose truth can be demonstrated and debated. Myth is not to be confused with fiction, a story which does not pretend to be true, but has entertainment or other value.

Many myths deal with the adventures of superhuman beings. Here we will also use the term to cover theories and descriptions of geological events and chemical reactions. The manner in which an account is presented will determine whether we consider it to be science or mythology. The person who presents a myth presumes that it is true, and does not consider any alternative explanation. He may produce evidence to support the myth, but he would continue to believe in it even if no evidence existed, or if it pointed in the other direction. For example, a person may believe that his birthday will bring him good luck. If he found money in the street on that day, he would cite that as proof of his luck. On the other hand, if he sprained his ankle on his birthday, he might ignore the connection or suppose that he would have broken his leg instead, if the accident had occurred on any other day.

An idea or account need not be wrong just because it is presented as a myth. In this book, though, we seek answers from science, not mythology. The mere statement that something is true need not be considered evidence in its favor no matter how many voices join in the chorus.

Myths are told wherever humans exist, and they satisfy many needs. They are often a vital part of religion, though religions have many additional elements, such as rituals, behavior codes, and value systems. Myths are important cultural institutions as well, and give meaning to the rules and traditions of a society. Beyond that, they provide necessary psychological support to individual human beings.

Consider the possible plight of a primitive farmer. He may have worked long hours in his field, tended in a caring way to the needs of his family, and supported the traditions of his community. He then sees his crop swept away by floods, his house destroyed by lightning, and his family and neighbors devastated by plague. He might give in to despair, feeling that all effort is useless, that he cannot control events, and that the world is a frightening and terrible place.

On the other hand, if he can feel that he has somehow offended the gods and they have punished him, some dignity is restored. The outside events resulted from his actions, and he may learn to control them to better effect. He can understand the anger of other humans and learn to cope with it. If nature is given human qualities, he can relate to it as well.

Even in cases where a human being may feel blameless, and dreadful events make no sense, myths can help mend the harm and offer hope. Many of us had parents who seemed all-wise and powerful but subjected us to painful experiences without any apparent reason. We trusted that things would work out well in the long run. Natural events, seen in the same light, are easier to endure. The famous biblical tale of Job describes an upright man with seven sons, three daughters, and numerous domestic animals. To test Job's faith, the Lord permits Satan to destroy Job's family and livestock and to afflict Job with boils. After much self-searching, Job keeps his faith and is rewarded. He rears a new family, again with seven sons and three daughters, and prospers, with a herd double its previous size.

Mythical accounts and religious beliefs give humans enormous comfort in the face of adversity. To be effective, however, they must be held firmly, and not subject to doubt. Unresolved questions, unclear answers, and changing views work in the opposite direction. They raise anxiety in us about our safety and destiny. For many of us, any firm answer which conveys a sense of purpose is better than no answer at all.

Creation Myths

Throughout history, myths have provided answers to the central questions concerning our existence, including the origin of humanity, all of life, and the universe. Generally, these subjects are linked together. Creation myths come from virtually all cultures, and collections such as *Sun Songs* by Raymond Van Over emphasize the many common themes. Not only similarities but also differences exist among the various myths. One variation is particularly relevant to this book, as it extends beyond mythology to conflicts that divide science and also separate science from mythology. At its basic level this dispute concerns whether creation stems from an individual being or from the universe as a whole.

In many creation myths, the existence of everything flows from the actions of an all-powerful creator. In this respect a Samoan creation myth resembles our own Bible. It begins: "The god Tagaloa lived in the far spaces. He created all things, He was alone, there was no heaven, no earth. He was alone and wandered about in space."

The origin of this first powerful being is rarely questioned in accounts of this type. He had no beginning, and has existed eternally. Often, he has human form, but exceptions exist. A myth from the Sia Indians of New Mexico, for example, relates: "In the beginning, long, long ago, there was but one being in the lower world. This was the spider Sussistinako. At that time there were no other insects, no birds, animals, or any other living beings." In this account, the spider then creates all other living creatures.

Other myths exist in which the creator has less purpose and power—for example, the tale of Father Raven in the first chapter. These powers may be even further restricted and barely exceed our own. The tale of the Old Ones in the first chapter would make an admirable myth, with a rather limited creator.

A creator with limited power and who is younger than the universe represents an intermediate rather than an ultimate solution to the search for origins. We would then ask what force was originally responsible for the start of life. An alternative answer invokes the germinal power of the universe itself as the source of life. This answer appears in myths from various cultures. Dr. Heinrich Brugsch summarized a compilation of Egyptian myths in the following way, as cited in Van Over's book:

In the beginning there existed neither heaven nor earth, and nothing existed except the boundless mass of primeval water which was shrouded in darkness, and which contained within it the germs and beginnings, male and female, of everything which was to be in the future world. The divine primeval spirit, which formed an essential part of the primeval matter, felt within itself the desire to begin the work of Creation, and its word woke to life the world, the form and shape of which it had already depicted within itself.

The *Rig-Veda* of India speaks similarly of a primal unknowable chaos from which the form of things arose. The Chinese philosopy of Lao-tzu tells of the *Tao*, a quiescence without form, which by spontaneous action created all things. This ancient alternative tradition of mythology has, in our times, reappeared in the heart of the scientific approach to the subject: life arises from preexisting matter that is unorganized, but has within it the potential to create the forms we know. On this note we will leave mythology and consider the very different means that science has employed to reach this same position.

The Rules of the Game

Science derives from *logos* rather than *mythos*. It uses a different approach to understanding the world around us. Those who wish quick satisfying answers are better served by mythology. The Skeptic, had he been so inclined, could have ended his search on the first day and looked no further. Many people choose to accept unified belief systems which provide answers to the major problems concerning life, and save themselves the bother of further inquiry. Their needs are met by mythology. In science, on the other hand, the method by which an answer is sought is more important than the nature of the solution. Questions need not be answered at all, or answers may be provided and later rejected, displaced by a newer theory.

The primitive farmer whose life has been devastated by floods, lightning, and plague will receive little comfort by taking up a scientific study of these matters. Eventually, however, he or his descendants will learn to build dams, erect lightning rods, and develop vaccines. Future calamities will be avoided. Even as the explanations offered by science alter, the technological innovations associated with them will endure and improve.

This visible progress of science differentiates it from many other human activities. The plays of Euripides, for instance, are still performed. The philosophy of Plato is still taught and debated. But the scientific theories of Aristotle are as dead as the gentleman himself, except to historians. Progress is possible in science (as it is in a baseball game) because theories, like teams, can lose. One mark of identification that a theory is scientific is the presence of some process by which it can be refuted in favor of another one. This takes place through observations and experiments made on the world around us.

The universe we inhabit and observe provides the ultimate source of authority in science. No statement in any text or word of any individual, however prominent he may be, supersedes it. Spontaneous generation was abandoned when experiments no longer favored it, despite its roster of illustrious supporters through the ages. Disputes in science are settled by making additional observations, not by debates or polls. But here, science differs from sports in that the final result need not come at once, as in the World Series. More often, it is a gradual process.

Baseball scores are definite; with rare exceptions, completed games are not repeated. In science, however, the basic set of data used to construct theories may shift and change, as mistakes are discovered. The amount of error that can enter in the course of making simple observations is much greater than nonscientists generally recognize.

The Abundance of Error

A popular maxim states, "I'll believe it when I see it with my own eyes." When I started work in my laboratory, I soon learned that I could not put complete trust even in my own eyes, let alone those of others.

We suffer in many ways in our perception, with a tendency to see what we have seen before, or wish to see. In a well-known series of experiments by J. S. Bruner and Leo Postman, subjects were shown a playing card for a very brief time, then asked to identify what they had observed. They did very well indeed when shown normal cards, but unusual cards were a different matter. A black four of hearts was almost always identified as black four of spades or a red four of hearts. Only when the card was shown repeatedly did the responses change. The subjects became confused; they understood that something was wrong. Finally, most of them recognized what they had seen. Others never got to that final point.

Errors in observation are not limited to untrained observers with very brief viewing periods. The noted astronomer Percival Lowell was convinced for many years at the turn of this century that an extensive canal network covered the surface of Mars. He constructed an elaborate set of fantasies about the inhabitants who had built the canals. Lowell named the various canals and prepared detailed maps which showed interconnecting straight lines that extended for thousands of miles. Decades later, when orbiting spacecraft photographed the surface of Mars in detail, no such canals were seen, nor any features corresponding even roughly to them in shape or location. Lowell had been the victim of an optical illusion caused when disconnected irregular features are examined at the limit of human visual perception. Under these conditions such patterns are seen as straight lines.

If we cannot accept the evidence of our senses, how are we to proceed? Even photographs, meter readings, and digital electronic displays must eventually be read and interpreted with our eyes. We must go ahead anyway, while keeping in mind that any single observation or set of observations may be wrong. The more surprising a finding, the more reason to distrust it. When I determine the temperature at which a new chemical substance melts and find it to be a routine number, I am inclined to accept it. If, on the other hand, I saw the substance rise slowly out of its test tube and hang suspended in the air, I would not conclude that it had learned to fly. I would either doubt the evidence of my senses or seek some other, more conventional explanation for what I had seen. I wouldn't disregard the observation, but I would want to get some confirmation from others. A wise philosopher said some centuries ago that for him to accept a miracle it would be necessary that the evidence supporting it be so impressive that the falsity of this evidence would be an even greater miracle.

To Persuade, Publish

The backbone of the scientific process is publication, the preparation of a full report of experiments with enough detail to allow another researcher to repeat it, if necessary. Ideally, publication should take place in a respected professional journal, one that uses referees. These individuals are experienced scientists familiar with the particular area, who may spot errors in the way an experiment was carried out or see that the conclusion does not follow from the data.

My wife, Sandy, once told me of a novel occasion where a referee performed his job directly at the site of an experiment rather than later, when the data were reported. Sandy is an academic psychologist. An older colleague, an expert in Soviet psychology, had informed her of recent exciting developments from that country. The Russians had reported that certain gifted subjects had the ability to perceive colors with their fingertips. Finally one such individual was located in the New York area. Sandy's colleague told her of the sad events that occurred when this subject was tested.

The woman had been blindfolded and seated at a table. Cards of various colors were put into her hands. She ran her fingers over them and, after a time, correctly named the color of each one. The demonstration convinced all those present, except for one skeptic. He wanted an appropriate referee for the procedure, and called upon a professional magician. The magician quickly pronounced his judgment: "She's peeking."

The woman was of good character, and nobody had expected that she would cheat. In fact, she herself seemed unaware of what she was doing. She had concentrated on her fingertips, and in doing so strained her facial muscles. Eventually, a flash of color appeared. Perhaps she imagined that she perceived it within her brain. In fact, a tiny amount of light had penetrated up under the blindfold. When the experiment was repeated under conditions in which vision was impossible, the effect disappeared.

On occasion, stories appear in the newspapers about researchers who report fraudulent scientific results. One such case involved a scientist who painted spots on the backs of mice, to simulate the effect he wanted. Scientists are human, and such events will occur. The threat of eventual disclosure seems enough to hold such incidents to a reasonable level. Much more common are inadvertent errors, in which a researcher sees a result he desires and rushes to embrace it, without pausing to take sufficient precautions against mistakes. Ideally in science, the individual who makes an exciting discovery should play the devil's advocate. He himself should take the most skeptical view of the results, and make every reasonable effort to find a less exciting explanation for it. Only after such efforts have failed should he publicize it. I hesitate to label this rule as essential, for it is honored about as much as the posted speed limit on major American superhighways. When I do see that a study has been carried out this way, I regard it as a label of superior quality. The absence of this attitude does the opposite, it raises a warning flag: let the reader beware, these results may be rubbish.

Publications Can Perish

Not all errors can be nipped in the bud before publication, as in the fingertip color vision experiments. Many pass the scrutiny of referees through oversight, or because insufficient or incorrect information was provided in the manuscript. I can recall one vivid case from my own experience.

I had been annoyed by a highly regarded professor at the California Institute of Technology. He had published two papers, in the most prestigious journal of chemistry, that had great implications for my field. The papers were superbly respectable, bristling with tables, graphs, and endless calculations. There was a problem, though. Earlier workers had studied the same question by other methods and reached opposite conclusions. The earlier work was very carefully done and seemed to have no flaw in it. In pushing his theory, the Cal Tech professor had made no reference to those earlier studies.

Soon thereafter, I got my chance for a strike at him. The setting was glorious: an Ivy League campus early in May. The sun shone brightly on the many flowering trees, creating a beautiful backdrop ideal for informal chats during the breaks. The scientists got none of it, however. The organizer, an aggressive, burly young man, allowed the talks to spill shamelessly over their time limits. The conference ran from breakfast until late at night in a darkened windowless room. Finally, on the second day, the Cal Tech professor spoke.

He presented his published data, which was greeted with great enthusiasm. "This is the most exciting news we've heard at the conference," boomed the organizer. At last, I got the floor. "What about all the earlier work that contradicts your own?" I asked, giving an account of it. He looked at me as though I had asked him for the name of the mayor of Shanghai. He shrugged, said that he hadn't considered the matter, and turned to another questioner.

"But then your data are quite possibly all wrong!" I blurted out. No one heeded me. I looked wildly around the room for assistance. There was one other person at the conference, deeply respected, who would know the old work as well as I did. He was a clever Scotsman who had helped found this area of research. But he was nowhere in sight.

Five minutes later I found him. "Why didn't you support me, Dan?" "Ach," he said, "I've just come from the men's room. Did anything interesting happen?"

The book and film *The Godfather* made it clear that revenge can still be sweet even when it comes "cold," that is, months or years later. So it was in this case. One year later, a retraction came from Cal Tech. Duplication of the results in other laboratories, and at Cal Tech, had failed. The papers, the mammoth piles of data, were all nonsense, the products of a childish experimental error that had been denied in the report. The professor apologized to the scientific community for the fiasco he had caused. The conference members had missed both the truth and the beauty of the campus on that spring afternoon.

Scientific papers must be regarded in the same way as we look at a new word in a crossword puzzle. When it fits in well with the words already present, it is likely to be correct. In a case where it contradicts earlier entries, we simply cannot write over the earlier ones. We must remove them and find alternatives that fit with the new word that we favor. We have no such problems when the new word is placed in an empty area of the puzzle. It is wise to be careful just the same, and write it in lightly in pencil. If we assume too firmly, in crossword puzzles or science, that new findings are correct, our assumption may block further progress in an area.

The treatment of science by the media and the public often does not reflect this caution. Unpublished results reported at a meeting are considered as fact. Published works are regarded as if they were graven on tablets of stone. Statements such as "a proven scientific fact" have become common currency in advertising and arguments alike. This phrase does not reflect the nature of science but rather displays an unfulfilled hunger for mythology. We scientists share this hunger particularly when our own efforts are responsible for the production of the myth. We get excited and feel gratified when we have a flash of insight or some new effect turns up in our laboratory. When one or two bits of confirmation fall into place, our attitude firms up: we now have the Truth. This feeling then prejudices our future efforts. We cannot avoid this human tendency; the best we can do is to be aware of it, and guard against it when it appears.

The Art of Theory Construction

The hazard that our data may be incorrect is one of several involved in the pursuit of science. Another horrid one is that our observations may be trivial, as I can show in an example: As I write this, I am looking out the window of my study at the leafy trees that surround my home. They provide many opportunities for taking data. I could count the number of trees on the property, and the number of leaves on each tree. This work would be tedious, time-consuming, and subject to error unless done carefully. Such qualities would even cause some scientists to look upon it with favor. Unfortunately, it would be of no interest; no theory would emerge from the numbers. If I counted the leaves each day, however, and plotted this count against time, I would "discover" the response of trees to the seasons. This time, I would have found an important effect. Unfortunately, the news of it has leaked out earlier. Again, my efforts would be pointless.

The creative scientists are those who collect data that matter, see the important connections, and draw the right conclusions. No systematic guidelines exist for this process, and pitfalls abound. As an example, consider the man who got drunk on gin and tonic on Monday, vodka and tonic on Tuesday, and rum and tonic on Wednesday. What caused his drunkenness?

If we knew nothing about alcoholic beverages, our first inference would be obvious: the common factor, tonic, caused the drunkenness. The deduction may be incorrect, but it is scientific. We can use it to make predictions which can be tested. Our experiment almost suggests itself. Let the man drink tonic alone. When he does this and remains sober, we would learn that our first idea was not correct.

Incorrect theories are seldom abandoned immediately in the real world. Efforts are made first to save them by modifying them. In the above case we might now suppose that tonic produces drunkenness only when it is diluted by some other liquid. Tonic and ginger ale could be tested next. When this failed, additional qualifications might be added. Eventually a new approach might emerge. By some flash of insight, we might conclude that the rum, gin, and vodka each could cause drunkenness, and that tonic played no role in the process. The stage would be set for a critical experiment. Our willing subject would attempt to get drunk on each of these beverages without tonic. This time, he would succeed.

A crucial experiment of this type represents the scientific equivalent of the face-to-face confrontation in a prizefight. No winner will emerge from a debate between conflicting mythologies, but a victor is expected when a theory is tested in science. The winner, of course, does not become the all-time champion. New contenders may enter the ring at any time. The new theory, that drunkenness can be caused by imbibing any of three liquids, gin, rum, or vodka, would face problems on Thursday night when the same man got sloshed by drinking whiskey and soda.

Eventually a modified theory might emerge, which contained an exhaustive list of intoxicating beverages. New ones, upon discovery, would simply be added to the list. One day a chemist might notice that all of them contained ethyl alcohol and write a simpler statement: beverages containing ethyl alcohol cause drunkenness. This summary, and the earlier one, would correctly describe all of the data. Which one should be used? A scientific rule covers this case. The simpler statement of the two will be the one adopted. The person who applies this principle is said to have applied Ockham's razor, a reference to William Ockham, a fourteenth-century theologian and philosopher.

All of the above theories, right and wrong, simple and complex, fall within science, as they can be proven wrong. Consider by comparison the following statement. "Drunkenness occurs whenever the god Bacchus chooses to shoot someone with an arrow. The state lasts until the arrow falls out. Neither Bacchus nor his arrows can be detected in any other way." I can form some wonderful images as I read this account. If I believed it, I could live with less guilt. Bacchus would be to blame, not I, when I got drunk. Alas, the statement is mythology, not science. No way exists to negate it, prove it wrong. Bacchus produces drunkenness. Drunkenness is the work of Bacchus. The circle is impenetrable. A person who introduced this as science would definitely be running from home plate to third base.

Semmelweis and Childbed Fever

The above cases are whimsical. I would like to provide a more vivid account of the scientific process by retelling the tale of Ignaz Semmelweis, who devised important preventive measures to control puerperal fever, also called childbed fever.

Semmelweis was a Hungarian physician who served at a hospital in Vienna in the 1840s. Two obstetrical clinics in this hospital differed sharply in their rates of mortality in childbirth caused by the above disease. No specific theory existed to describe the cause of the disease or the differences in rates, only empty generalizations—for example, that it was due to "atmospheric-cosmic-telluric" influences. This description, which covered sky, universe, and earth, was so vague as to suggest no tests, and was unscientific and useless. Semmelweis chose rather to observe the two clinics closely.

Medical students received obstetrical training in the first clinic, the one with the high mortality, while midwives were employed in the other one. Perhaps the students were clumsier, and inflicted injuries by rough handling during their examinations? Inspection of the patients for such injuries revealed no significant differences. Another possibility then emerged. Women in the second clinic were delivered while lying on their sides, those in the first clinic on their backs. With some difficulty, the students in the first clinic were induced to adopt the side position. No change in mortality appeared.

A psychological explanation was considered. The first clinic lay adjacent to a sick ward where a priest was often summoned to administer last rites. An orderly with a ringing bell preceded the priest. Both of them passed through the first clinic, but not the second one, on their route. Did this grim and noisy spectacle frighten and demoralize the expectant mothers, reducing their resistance to illness? The priest was rerouted, but mortality rates did not change. Many other factors also were tested, to no avail.

A key observation was made by accident. Jakob Kolletschka, a colleague of Semmelweis, received a puncture wound in his finger while performing an autopsy. He died, with symptoms resembling those of childbed fever. Semmelweis decided that "cadaveric particles" introduced into the bloodstream of his colleague had caused the illness, and jumped to the conclusion that the women in the delivery room had suffered a similar fate. Medical students performed autopsies, washed their hands superficially, then went to the first clinic to examine the patients, and infected them. The midwives in the second clinic performed no dissections, and caused no disease.

Semmelweis then required that all students wash their hands in a solution of chlorinated lime before entering the maternity ward. This substance was sufficient to remove cadaver odor from their hands, and presumably destroyed the particles. Within two months, the mortality rate in the first clinic had fallen to a fraction of its previous level. Many lives were saved.

This satisfactory performance did not confirm all the details of the new theory. An unfortunate incident led to its modification. Eleven patients died at the same time of childbed fever. No cadaver was involved, but the epidemic was traced to another source. One patient in the same ward had suffered from a "festering cervical cancer." The medical personnel who had examined her then proceeded to examine other patients in the same ward, without pausing to wash their hands in chlorinated lime. It was then recognized that not only material from corpses but also "putrid matter derived from living organisms" could cause the disease. Improved procedures were adopted and additional lives were undoubtedly saved. Despite this success, the actual cause of the disease, infection by microorganisms, was not yet understood.

The shortcomings of the theory as well as opposition based on political grounds delayed the acceptance of Semmelweis's disinfectant procedures. Ironically, he died of an infected wound, as had his colleague Kolletschka, before his triumph was realized.

The Parade of Paradigms

The saga of Semmelweis illustrates how particular ideas, with predictive success, may later be discarded in favor of more effective ones. This fate applies not only to individual theories, but to much broader explanatory concepts that link an entire field. Such concepts have been termed "paradigms" by the philosopher Thomas Kuhn in his germinal book, *The Structure of Scientific Revolutions*.

Semmelweis attempted to combat disease at a time when the relevant science was in a pre-paradigm state. Much important data concerning mortality had been collected, but no unifying concept was at hand to explain it. Science, lacking a paradigm, can be a haphazard affair. Data are collected essentially at random. Different competing schools spring up, and each interprets the information according to its own presumptions. The adherents of each school generally ignore the findings of the other schools. New speculations appear continually. (A speculation is a scientific explanation that extends far beyond the available data. It can be tested in principle, but generally it is not convenient to do so at the present time. Crick's idea that life began on earth when bacteria arrived by spacecraft is a good example of a speculation.)

Pre-paradigm areas of science usually excite the general public, but frustrate the scientists that work in them. The questions concerning the molecular basis of aging and consciousness or the existence and nature of life elsewhere in the universe are areas of this type.

Eventually, as a field matures, one school of thought triumphs. Its particular way of interpreting the data proves more effective, and makes better predictions than the others. The winner becomes established as the ruling paradigm. The atomic theory of matter, Darwinian evolution, and the molecular basis of heredity, among others, fall into this category. A paradigm, once established, dominates thinking in its area. New students are initiated into the field by studying it. Books and articles in the area, previously comprehensible to the layman, now assume detailed knowledge of the paradigm and become incomprehensible to the general public. Above all, a burst of new scientific activity takes place.

A new paradigm will provide only a broad outline for an area. Details must be filled in. The consequences of the paradigm must be explored in depth. Results which do not fit the picture must be checked, and brought within the structure if possible. Possible extensions of the paradigm to adjacent areas must be attempted. Such activity, confirming the existing picture, is termed "normal science" by Kuhn. Most of the results may hold little interest for the public in general, but this type of work brings satisfaction to scientists. Experiments, when done with skill, give results that make sense. Another piece is put into a picture puzzle whose overall content is clear. The best results draw the appreciation of almost all of the workers in that area.

On occasion, such intensive study of an area will turn up new anomalies, fresh pieces that do not fit in. Errors of the type we have discussed will account for many of them. Every healthy area of science will have a supply of such anomalies. (They provide convenient problems for Ph.D. theses.) Gradually, they are resolved and new ones take their place. But occasionally the anomalies do not give way. As attempts are made to resolve them, they multiply and become more obvious. Eventually, they threaten the paradigm itself.

At this point a feeling of crisis and uncertainty enters the area; the practitioners feel malaise. This anxiety arises from the emotional nature of the scientists involved rather than from any threat to the technical accomplishments of the field. Unpredictability and uncertainty have entered the tidy world of the paradigm, which in its secure state had fulfilled many of the functions of a myth. Like the heretic who is not welcome in church, the scientist who disputes the ruling paradigm is not embraced by his colleagues.

In some cases, the troubles multiply until the paradigm itself topples, displaced by another one. A scientific revolution has taken place. One such case was the replacement of the earth-centered astronomy of Ptolemy by the view of Copernicus, in which the planets and earth revolve around the sun. In other instances, a paradigm may collapse of its own weight with no successor, and the pre-paradigm situation returns for a time. Spontaneous generation represents such a case.

Accounts of the development of science assume a gradual accumulation of understanding, a smooth climb up the ladder of knowledge as history progresses. Kuhn views the process as a series of discontinuous episodes, the rise and fall of paradigms. The history of the origin-of-life question is best viewed in this context. Spontaneous generation governed the area for millennia. It sagged in the eighteenth century but did not collapse fully until the 1860s, when an important series of experiments were performed by Louis Pasteur. A period of confusion followed, until a new paradigm emerged over the period 1922–1953. It has been named the Oparin-Haldane hypothesis after its founders, Alexander I. Oparin and J. B. S. Haldane. This theory prevails today, but its grip has grown infirm. Anomalies have appeared and now threaten the basic structure. New speculations, candidates for the role of future paradigm, have emerged. The outcome is not settled, but we will better appreciate the present difficulties when we have digested the example of the past.

Spontaneous Generation: Paradigm Lost

The term "spontaneous generation" has been applied in a number of ways. We will adopt here the definition provided by the historian John Farley. It is the belief that "some living entities may arise suddenly by chance from matter independently of any parents." This idea reflects the experience of many observers back to the time of ancient China, Greece, and Babylon.

I can supplement their observations with one of my own. I toured the Galápagos Islands recently to see the sites that inspired Charles Darwin and furnished so much data for his later theory. My fellow voyagers and I explored one island, Fernandina, which contains vast lava fields, the residue of intermittent eruptions over several centuries. Little life was to be seen in this enormous stretch of curved and jagged dark stone, which extended from the mountains to the sea. The most notable exceptions were obvious only at close range, for their color and shape blended with the rocks. Tiny gray-black lava lizards darted about in many places. Larger spiny black reptilian forms, the marine iguanas, basked near the sea. So well did they fit their background that I imagined them to be derived from the lava, the product of spontaneous generation. Alexander Oparin had earlier summarized this temptation: "Whenever man had met with the unexpected and exuberant appearance of living things, he has regarded it as an instance of the spontaneous generation of life."

The collapse of the paradigm of spontaneous generation began when man substituted active experiment for passive observation. An Italian physician, Francesco Redi, was among the first to provide cause for doubt, in the seventeenth century. Redi stored some fresh-killed snake meat in a container open to the air. As many others had observed, small white worms, maggots, appeared in the meat after several days. Redi removed several of them and placed them in a separate vessel. After further time passed, each developed into a fly. They had not been worms, but insect larvae.

He now repeated the experiment, but covered the vessels in which the meat was stored with gauze. The mesh was so fine that flies could not approach the meat. No maggots developed inside, but insect eggs appeared on the gauze. The protective cover was now removed, and maggots appeared on the meat in due course. Their source had been shown to be flies, rather than spontaneous generation. The idea had been negated in this particular case, yet the principle survived. Redi himself believed that spontaneous generation could occur in different circumstances.

One particular case accepted by many scientists was the spontaneous generation of microbes. These "animalcules" had been discovered by Antonie van Leeuwenhoek, a contemporary of Redi, in his pioneering investigations with the microscope. John Tuberville Needham, a Welsh naturalist and Jesuit priest of the eighteenth century, maintained that he had observed spontaneous generation of these tiny creatures in various nutrient broths that he had prepared. Needham boiled his broths, to kill microorganisms that already were present, then sealed the flasks, sometimes in an airtight manner. After sealing the vessels, he heated them in hot ashes to sterilize the air within them. No precautions were neglected, he claimed. In all cases, "animalcules" appeared within the flasks after several days.

Needham's views were opposed by another scientist-priest, the Italian Lazzaro Spallanzani, who ran the same type of experiment with greater care. Spallanzani first sealed all his vessels in an airtight manner, then heated them for longer periods, to ensure sterilization. In hundreds of such experiments using many recipes for the broth, no microbes appeared. He concluded that Needham had either taken insufficient precautions in sealing his vessels, or had not heated them long enough.

Needham did not thank Spallanzani for the elegant negation of this theory. Perhaps his vocation had not prepared him for the role of devil's advocate. Rather, Needham adjusted his theory to meet the new circumstances. He felt that his broths, which he called infusions, did have the power to create life, but that their vitality could be destroyed by rough treatment, à la Spallanzani. Needham's own words can be cited: "But from the method of treatment by which he [Spallanzani] had tortured his nineteen vegetable infusions, it is plain that he has greatly weakened, or perhaps entirely destroyed, the vegetative force of the infused substances." It was not obvious at that point how a critical test could be performed, and the controversy struggled on until the time of Louis Pasteur. Pasteur won a prize in 1862 from the French Academy of Sciences for his experiments related to spontaneous generation. A colleague, J. B. Dumas, had cautioned him on entry into the study of the origin of life: "I would not advise anyone to spend too long on the subject." Pasteur did well for the time he invested in it, but, more than a century later, I have received similar advice myself.

Pasteur demonstrated that supposed cases of spontaneous generation were due to contamination of broths by microorganisms carried on dust particles in the air. In key experiments he used swan-neck flasks, so named because long S-shaped necks connected them to the outside air. The broths within were sterilized by heat, and remained sterile. Dust particles carrying bacteria were trapped within the neck and could not reach the liquid. When the necks were removed, however, the broths yielded hosts of microbes within forty-eight hours. The initial absence of bacteria in the sterilized broth was not due to loss of vegetative power, but rather to exclusion of microbes from the air.

Pasteur summarized his work with a triumphant lecture at the Sorbonne in 1864, which he concluded with this remark: "Never will the doctrine of spontaneous generation recover from the mortal blow of this simple experiment."

The blow may have been a mortal one, but the victim took some time to pass away. A scientific maxim states that discredited theories expire not by the rapid conversion of their followers, but only after the last adherents have died off. The final survivor from that time supporting spontaneous generation was an English scientist, Henry C. Bastian. He had discovered that hay infusions contain unusually heat-resistant spores. Much longer heating times were needed to destroy them. He did not interpret his results in this manner, but saw them as proof of spontaneous generation. In the 1870s he engaged in bitter debates with members of the French Academy. He continued alone in support of his position until his death in 1915.

Bastian's example demonstrates the hold that a paradigm or a theory of one's own can have on the mind of man. The skeptical position more suited to science is abandoned, and the idea takes on the attributes of a myth. We shall encounter this behavior again as we search for the origin of life. Before we can consider theories more modern than spontaneous generation, however, we will have to pause to review some of the fundamental information that has been gained by science concerning the nature of life and its history on this planet.

two

Two Stains on a Rock

It is easy to look at a dog and a rock and decide that one is alive and the other is not. It is much harder to compare two stains on a rock and come to the same conclusion. Yet one stain may be mineral, quite similar to the remainder of the rock, while the other may be a primitive plant form (a lichen), composed of the same chemicals as the dog.

The nature of one group of rock stains, as yet unidentified, has immense importance. In 1976 two unmanned landers of the Viking project arrived on the surface of Mars and sought by various means to determine whether life was present. The cameras that they carried provided the most direct method for detection of life. The advance information concerning the surface of Mars was so sketchy and incomplete that even the presence of animals as large as polar bears could not be excluded.

The cameras revealed nothing that moved, nor any features that testified to the obvious presence of life. Dr. Gilbert Levin, a member of the Viking investigator team, was not easily discouraged, however, and scanned the photographs with great care. He discovered that rocks near one of the landers carried green patches that bore close resemblance to lichen on earth. Lichen, which are actually a type of marriage between algae and fungi, are among the most adaptable life forms on earth. They can survive in cold, arid places such as mountaintops and in the Antarctic. They lie dormant when conditions are poor, and show a burst of activity when sunlight and moisture return. If any known life form were to be found on Mars, lichen is a likely candidate.

Unfortunately, the investigation remains suspended at this point, because no samples of the stains could be retrieved for analysis. We must wait until the exploration of the surface of Mars is resumed at some future date to learn the nature of these patches. If we could somehow transport a sample of this material to earth, there would be little problem in identifying it. Under the microscope, lichen show characteristic cells and filaments, while minerals ordinarily present a very different appearance. Chemical analysis would provide even more definite results. Certain atoms and molecules are characteristic of living things on earth, and very different ones are present in rocks. These tests are derived from our long experience with both lichens and minerals, but do not catch the essential difference between living creatures and nonliving matter. Yet it is this very difference that we must explore fully if we are to explain how one may have arisen from the other.

Let us return to our first comparison of dog and rock and consider organization rather than animation. The body of the dog can be divided into various components: a head, legs, a trunk, and a tail. Rocks generally do not display any such obvious organization. Even if we could locate an irregular specimen, with apparent subdivisions, this shape would be accidental. Other dogs will have the same parts as the first one we observed, but rocks will differ from one another.

The inside of a dog is organized as well. The many different organs have their own particular place within it. This organization extends down to smaller and smaller components. Organs are constructed of tissues, which in turn are made of cells. The cells themselves are made of characteristic parts. No such hierarchy of well-defined levels of organization exists within rocks.

The theory of evolution indicates that the higher levels in the organization of life arose after the lower ones. We shall see that the oldest cells detected in fossil form were simple ones. Generally it is believed that more complex cells came later in evolution, and organisms made of many cells later yet.

The origin of life, then, involved the organization of the lowest levels: molecules and cell components. We must learn how life functions at these levels today before we can inquire how this situation arose for the first time. We shall pause to explore this sub-microscopic world.

The World of COSMEL

It is difficult, but not impossible, to visualize how the size of a cell or atom compares to objects in our everyday world. We will do this with the aid of an imaginary device called COSMEL, the cosmic orders of magnitude elevator. While an ordinary elevator takes us to higher or lower floors, COSMEL appears to enlarge or shrink us in size. We enter at level 0, which marks the ground floor, and may push buttons marked from 1 to 25 to move in an upward direction or from -1 to -15 to descend to lower levels. Each positive number increases our apparent size by a factor of 10 over the one just below it, while each negative number will decrease our size to a similar extent, compared to the one just above it.

If we were to push the number 1 and ascend to the first level, for example, we would seem to be 10 times our normal size. Thus if our height normally was 180 centimeters (6 feet), we would step out into a world in which we appeared to be 18 meters (60 feet) tall. People would come up to our ankles, and trees would resemble shrubs. If we had pushed the button marked 2 instead, we would emerge to find that we stood at the height of skyscrapers. Those with a mathematical inclination may have noticed that the number on the button in COSMEL represents the power of 10 to which our apparent size has been multiplied. Thus on the second level we would be 10^2 or 100 times our normal height.

These sizes that we perceive are an illusion. The laws of nature do not permit us to exist in a wide range of sizes and conduct our business as usual, the Lilliputians and Brobdingnagians of Jonathan Swift notwithstanding. If our height increased 10 times, for example, our body surface would increase about 100 times, and our weight, the total amount of our flesh, 1,000 times. The heat produced by our bodily activities would go up in proportion to our weight, but we would have less surface area available to dispose of it. We would soon roast in our own heat. Before that time we would have crumpled to the ground. The strength of our legs would have increased 10 times in proportion to their cross section, an amount insufficient to enable them to support our weight.

COSMEL is best viewed as a series of models, designed cleverly to reflect how the world might appear if we could change our size. The levels could be imagined, or actually constructed on different floors of a museum, and be reached using an ordinary elevator with appropriate markings. In our search for the origin of life we will not require the levels above the ground floor, but rather the lower ones. To start our exploration, let us consider this very book. It has a size when closed of 23 by 15 by 3 centimeters (9 by 6 by 1¼ inches). Let us push the -1 button of COSMEL and examine it again. It has become a slab with roughly the dimensions of a king-size bed.

For our remaining voyages we will keep the book open to this page. Let us focus our attention on any letter *i* on this page and push the -3 button. Our size will seem one-thousandth of its normal one when we emerge from the elevator, the same height as the letter *i*, excluding the dot. We would just fit, if we lay down on the *i*. The dot itself would be a black spot 30 centimeters (1 foot) wide, the size of the top of a wastebasket. The entire page on which we stood would occupy an area 3 city blocks long and 2 blocks wide, enough to form a large public square. If we walked to the edge and looked down, we would be perched on top of a cliff the height of a 6-story building. The side of the cliff would resemble an end-on view of a stack of rugs. Each "rug," a page of this book, would be about 6 centimeters (2½ inches) thick. The surface of the page we were standing on would also be ruglike rather than smooth, with woven strands, channels, and hollows easily visible.

We are not making this excursion to study the book publisher's art, but as part of a quest for the origin of life. To sharpen our understanding of life, a model of a simple organism, a paramecium, has been placed on the dot of the *i*. In the real world, paramecia inhabit freshwater ponds, rather than book pages. In our model, it is roughly the size of our hand and stretches about halfway across the dot. On closer inspection, we can see that it has a squat cigar shape and is covered with hundreds of short hairlike structures, called cilia. A pore, used for sweeping in food, appears on one side of the creature.

The meals of the paramecium often consist of bacteria, tiny organisms that are among the smallest that live on this planet. Some have been provided in this model, placed near the "mouth" of the paramecium. They are quite tiny at the -3 level of COSMEL, about the size of a printed letter *o* in our everyday world.

The cell, a fluid-filled enclosure surrounded by a membrane surface, is a basic unit of biology. Our own bodies contain many trillions of them. The two creatures we have examined each consist of a single cell, despite their difference in size. Each of them would seem to have much more in common with the other than with us, yet a fundamental division in biology places the paramecium and us in a single group, termed the eukaryotes. We are not alone together, but share the group with almost every other familiar living thing from asparagus to zebras. The other class, the prokaryotes, contain primarily the bacteria and the blue-green algae. The basis for this division is the complexity of the construction of the individual cells. The features that unite us with a paramecium can be observed if we place a model of a typical human cell on the book page at the -3 level and arrange that both the paramecium and it are illuminated from inside, to reveal their contents.

The human cell would appear coin-sized, smaller than the paramecium, and would lack the cilia and mouth pore. Both cells, however, would reveal a prominent inner compartment, called the nucleus. In addition, both would contain a bewildering array of sacs, tubes, and other structures, called organelles. The models would reveal a great richness of inner detail, and many similarities.

The bacteria are too tiny to be examined at the -3 level, so we must take another trip with COSMEL. We push the -6 button and emerge on the dot of the *i*. We now appear one-millionth of our normal size, while the dot has expanded its dimensions at the -3 level by 1,000. It measures about 330 meters (⅕ mile) across, the size of a small lake, while the main part of the letter *i* now stretches 1¾ kilometers (1 mile). The edge of the page lies many kilometers away. We would not enjoy a trip to the edge, or even a short stroll around the perimeter of the dot, as the terrain has become quite rugged. Thick fibers of paper material, a chemical called cellulose, tower about us while crevasses and craters mark the ground. We will limit our exploration to our immediate vicinity, which contains a single bacterium. We are viewing the world from a bacterial perspective.

The bacterium resembles a rounded cylinder, 2 meters (6½ feet) long and 1 meter (3¼ feet) across. The paramecium, by contrast, would appear a monster, the size of a small warship. Six whiplike filaments, called flagella, project from the bacterium, each longer than its body, though no thicker than a finger. They are used to propel the creature.

We have equipped our model bacterium with internal illumination, which we can turn on to examine its contents. We should note in passing that the smaller objects within a real bacterium could not be examined using ordinary light, even under a microscope. Visible light cannot resolve objects below a certain size. Scientists have explored the fine structure of bacteria using a different type of illumination and a special instrument, the electron microscope.

Our model presents no such problems. We can see that a rigid netlike material, the cell wall, surrounds the bacterium, with a smoother layer, the cell membrane, beneath it. The flagella are connected to these coverings by a hooklike end which contains a series of rods and rings. The interior of the bacterium shows a number of complexities, but far fewer than those present in the paramecium or human cell. One of the organelles of the paramecium or human cell would fill up the entire volume of the bacterium. No nucleus is present in the bacterium either, but simpler items can be seen. Tiny spheres, the size of a quarter, are scattered throughout the fluid within the creature, and in some cases a number of them are linked together by a thread. These objects, called ribosomes, are universal to all cells. Another feature in our model is a structure attached to the inner side of the cell membrane, which resembles a string twisted in many loops around a central core. This object, the bacterial chromosome, contains a chemical called DNA.

These cell components, modest in size even when viewed at the -6 level of COSMEL, do not yet represent the lowest level of organization of life. They are constructed of particular kinds of molecules arranged in a highly specific way. These molecules in turn are made by connecting certain atoms in unique ways. We must start our study of life's organization, then, at the level of the atoms.

A Universe of Atoms

Religious fundamentalists contest the theory of evolution, and the Flat Earth Society disputes even the roundness of our planet, but no organization exists, to my knowledge, that opposes the atomic theory of matter. All those who follow science agree that atoms exist and determine the properties of matter, even though they are too small to be observed directly or with a microscope. We can see them in our COSMEL model, however.

We are at the -6 level, in which the dot of an *i* has been expanded to the size of a small lake. If we were to examine any object around us very carefully, a strand of cellulose from the paper, a splotch of ink, or a bacterial flagellum, we would find it to be granular in consistency, like beach sand or a photograph reproduced in a newspaper. The granules, barely visible to our eyes, would be atoms. It would be arduous to count the number present in the bacterium, for it contains about 200 million of them. The differences in the types of atoms present, and their arrangement, are responsible for the properties that distinguish paper, ink, and a bacterium from one another. To explore these differences, we will interrupt our trip with COSMEL and enlist the aid of another imaginary device, the Atom Grinder.

Unlike our magical elevator, the new machine has a counterpart in the real world. Chemists can take any substance and, after applying several procedures and using various instruments, determine the types of atoms present in it. We wish to speed up the process in this narrative, so we will use a device that chews up anything and quickly types out the relative amounts of atoms present to the nearest half percent.

Atoms play about the same role in an object as letters do in printed matter. A book printed in English generally uses about 70 to 80 symbols. There are 26 small letters, 26 capital letters, 10 numerals, and about a dozen common punctuation marks. The universe contains more than a hundred types of atoms, some made artificially by man and so short-lived that they play little role. In the English language, the arrangement of symbols (we will call them letters from now on) into words is more important than the total amount of a particular letter present. This book, the current best-selling novel, and the Bible probably have a similar content of the letter *s* or *e*. If they should differ, this difference is unlikely to be related to the content of the books in any significant way. When real objects are examined, however, both the types of atoms present and their arrangement are significant. We can illustrate this now using the Atom Grinder.

To start, we will allow a stream of air to flow into the machine. It whirrs and chugs, then types out: nitrogen 77%, oxygen 21%, hydrogen 1%, and argon 0.4%, other atoms only in trace amounts. Only four of a hundred or so possible atoms are represented to any extent. The arrangement of atoms in the air is also simple. To illustrate this, we will again use the English language for comparison. Letters are organized into words, while atoms are united by chemical bonds into molecules. Words seldom exceed twenty letters, but molecules can have very large numbers of atoms. However, only simple molecules are present in the air in any amount, the equivalent of one-, two-, or three-letter words. Argon exists as isolated atoms. It belongs to a class called the noble or inert gases. They form no bonds and do not enter into molecules. Nitrogen atoms unite in pairs in the air, to form nitrogen molecules. Atoms of oxygen behave similarly. Hydrogen atoms exist in combination with oxygen, in a 2-to-1 ratio, forming a molecule whose chemical description is H_2O, but which we know better as water.

One further comment must be added, to explain the airiness of air. The various molecules in the air do not cluster together but separate widely from each other. As an analogy, think of a book with only a few scattered words on each page.

To continue our exploration, we next pour some water into the Atom Grinder. It promptly types out (to the nearest percent): hydrogen 67%, oxygen 33%. The composition of liquid water is the same as the water in the air. A piece of ice put into the grinder would give the same result. Solid and liquid water differ from the gaseous form in that their molecules are close together, rather than separated as in the air. To explain the difference between solid and liquid, however, we had better change the metaphor and consider a group of individuals. The liquid state resembles a crowded dance floor where dancers move about and push past each other. To visualize the solid state, think of a filled theater, where individuals are seated close together but remain in place.

We will next examine a sample of bacteria to see which atoms are used in constructing life. Imagine that we collect a supply of them in a pond, then shove them into the Atom Grinder. After the machine has chewed them up, it reports: hydrogen 61%, oxygen 27%, carbon 8%, nitrogen 2.5%; many other elements are present, but none to the extent of 0.5% or more. (The word "element" is sometimes used rather than the phrase "type of atom." We can thus say that the universe contains over one hundred elements.) Bacteria are built primarily from four elements, three of which are abundant in air and water. The fourth one, carbon, is less prominent in our environment, but plays a vital role in the construction of life. Roughly 70 percent of the weight of a bacterium is water, and its overall composition is close to that of water. The remainder is a mixture of molecules of great complexity.

We have tested air, water, and life; now we shall try the earth. The center of our planet is believed to be largely molten iron, but we are not interested in that at this point. We wish instead to know about the composition of the crust, the surface rocks that interact with life. A rock has been selected whose composition reflects that of the crust as a whole, and we toss it into the grinder. We receive the longest analysis of any thus far: oxygen 48%, silicon 28%, aluminum 4.5%, calcium 3.5%, potassium 2.5%, magnesium 2%, and a number of others less than 1%.

Only oxygen appeared on our previous lists. Silicon is an important component of rocks, and plays somewhat the same structural role that carbon does in living matter. It binds to several atoms at once and forms molecules of very large size. The last four of the elements on the list are metals. Some are familiar to us from our kitchen or workshop, where we meet them in the free state (not bonded chemically to other substances). In this state, they are shiny, hard, heat-conducting materials, used in making tools, coins, weapons, and buildings. The metal atoms present in rocks most often are in a chemically combined form and show very different properties, just as rust may differ from iron.

Thus, in the Martian exploration described at the start of this chapter, determination of the types of atoms present in each stain and their relative amounts would have served to tell us whether it was lichen or a mineral. Unfortunately, the Viking lander did not have this capability.

Chemical analysis, then, is enough to detect the difference between a lichen and a mineral, or a dog and a rock, but it does not explain the difference between life and nonlife. We could readily assemble a mixture containing carbon, hydrogen, oxygen, and nitrogen, but it would not be alive. As we have seen, the last three kinds of atoms are abundant in a mixture of water and air. Carbon could readily be added to the combination. We could select the gas, carbon dioxide, which is familiar to us as the bubbles in club soda. Alternatively, we could add limestone, a type of mineral that contains much carbon, or use diamonds, which are made almost entirely of carbon. No mixture of air, water, and any of these substances would have the slightest resemblance to life. Nor would it help to add the various kinds of atoms present in life in trace amounts. Clearly, much more than the atomic composition is involved. We must examine the arrangement of atoms into molecules in greater detail. Carbon will be a particular focus of our attention, as we attempt to appreciate how life differs from nonlife at the lowest level of organization.

The properties of carbon have afforded a vast chemistry of

such complexity that an entire area, organic chemistry, has been devoted to the study of it. In comparison, all the other elements except carbon are included in a single alternative area of study called inorganic chemistry.

Carbon atoms have a marvelous ability to join with one another and with certain other atoms to form chains varying from two to millions of atoms in length. Such long chains are characteristic of many molecules important to life.

Yet once again, these circumstances do not define life. At one time, up to the early nineteenth century, it was thought that the division between organic and inorganic chemistry was the basis that separated living and nonliving matter. Now we know better. Certain meteorites, for example, contain a complex mixture of organic compounds, with chains of various length. Yet they do not contain life, nor is there any indication that they were ever in contact with life before they fell to earth. To complicate matters further, rocks also contain long chains of atoms, though of a different type. Silicon atoms prefer oxygen as a bonding partner, and together they form a group of atoms called silicate. In rocks, silicates link together to form long chains. The chemistry of these substances is quite complex, though less explored than that of carbon.

The essence of the difference between life and nonlife at the molecular level lies not in the presence of a particular feature, such as long chains of atoms, but rather in the organization, as well as the identity, of the molecules. To illustrate this, let us explore the interior of a grain of sand at the -8 level of COSMEL. In our model of this level, we will use small balls varying from the size of a grape to a Ping-Pong ball to represent atoms. Further, we will use different colors to help us recognize different atoms.

If we sampled a number of locations within the sand grain at random, we would generally find the same situation at each point: a three-dimensional network of alternating silicon and carbon atoms extending indefinitely in all directions. The common chemical name applied to this monotonous substance is quartz. We can also encounter monotony of this type in other substances. A diamond, for example, is made of a repetitive three-dimensional network of carbon atoms alone.

A very different experience would await us if we performed

the same exploration within a bacterium. Its membrane would appear as a thick wall, with various structures embedded within it. The interior of the wall would be made largely of two types of atoms, carbon and hydrogen, but oxygen atoms would embroider the external surfaces of the wall. A ribosome would appear as a roughly heart-shaped object, as wide in diameter as our own height. If we inspected our model more closely, we would notice that it was composed of two separate parts, each of which contained a number of complicated large molecules fitted together in a vast three-dimensional jigsaw puzzle.

Different but equally complex experiences would await us if we sampled other locations within the bacterium. The entire creature, at the -8 level, would rival an ocean liner in size and complexity. It would be an immense undertaking for us to explore all of its connections, atom by atom. Biochemists have been engaged in this very task for decades. They have not had the benefit of direct visualization, as we have on our imaginary trip, but have had to use indirect and laborious methods. Their task is only partly complete, but their achievements thus far have made possible the COSMEL model presented in this chapter. This is not the place to review the techniques that they used or the details of their discoveries, which would require many volumes for adequate presentation. We will, however, consider certain key features important to understanding the problems involved in the origin of life. To simplify our task, we will use the analogy of a book.

The Languages of Life

This book contains several hundred thousand letters, which carry much more information than they would if simply mixed together in an alphabet soup. At the first level of organization, the letters are collected together into words. Word formation alone does not convey the message. If we were presented with this book cut up into what we will call a word soup, we could learn that it was written in English and probably concerned some topic in science, but little more. The words must be further arranged into sentences, and the sentences into paragraphs and chapters which are placed into their proper order, before we can get the full intended message. This organization can be compared to that of a bacterium. Atoms are collected into molecules, as letters are into words. These molecules are connected to form much larger ones, called macromolecules, the equivalent of the combination of words forming very long, paragraph-sized sentences. Macromolecules combine to form larger structures, such as ribosomes, as sentences come together to create chapters. A ribosome is the equivalent of a very large chapter, as it has perhaps half as many atoms as there are letters in this book. The culmination of these layers of organization is the combination of cell components that form a bacterium, as chapters form a book.

But we must modify this analogy if we are to consider matters in greater detail. This book is written in one language, English. The bacterial "book," by contrast, is written in four different "languages" which are segregated into separate sentences or chapters. Their names, familiar to us from diets and works on popular science, are lipids (fats), carbohydrates, proteins, and nucleic acids.

Lipids: Protection and Energy Storage

Bacteria are not troubled by fat around their waistlines. They do not diet willingly. Lipids serve other purposes for them, only one of which need be mentioned here. They function as a skin, forming much of the barrier, called the cell membrane, that separates the contents of a cell from its outer environment. Their water-resistant nature permits them to fill this function, since neither water nor most water-soluble substances can pass through a lipid layer with ease.

This water-avoiding property of lipids results from an abundance of hydrogen atoms and a dearth of oxygen and nitrogen atoms. Chemists refer to this hydrogen-rich state as "reduced," while the opposite condition, oxygen-rich and hydrogen-poor, is termed "oxidized." Lipids are the most reduced of the common classes of molecules found in living things. We would be misled if we attempted to apply these concepts in our diets. A diet rich in lipids would not normally reduce our weight!

The small proportion of oxygen present in lipids is important to their biological function. It also distinguishes them from a group of mostly nonbiological substances made entirely of hydrogen and carbon, appropriately named the hydrocarbons. The difference between the two classes is underscored every so often when some unfortunate individual substitutes machine oil, a hydrocarbon mixture, for salad oil, sometimes with lethal results.

Hydrocarbons, unlike lipids, play little role in life today. The simplest hydrocarbon, methane (a component of natural gas), has particular importance in certain theories concerning the origin of life. In these accounts, methane was abundant in the atmosphere in the earth's early days, and supplied the carbon needed for the construction of molecules needed to start life. We shall return to this topic later.

Carbohydrates: Sweet and Strong

Carbohydrates comprise another important "language" of biochemistry. The individual "words" of this language are the sugars, while the "sentences" formed when these words are united are called polysaccharides. Sugars and polysaccharides together form the class called carbohydrates. The prefix "poly" simply means "many" (as in polygamy, many wives), while the ending "saccharide" signifies "sugar" or "sweet." Do not confuse it with the synthetic chemical saccharine, which, while sweet, is not a sugar.

The combination of individual sugars with larger units involves a principle opposite that involved in combining words into sentences. We add something, a space, when words are put together. We remove something when sugars or other biochemical "words," such as amino acids or nucleotides, are combined. This something is a molecule of water. If the parts should separate again at some later state, the water would be restored. Each time another unit is added to a growing chain of sugars, an additional molecule of water is split out. In forming a 100-unit polysaccharide chain, 99 molecules of water would be released.

Although thousands of sugars could be prepared in the laboratory, only a handful are important in biology. Often only one kind of sugar is linked together to form a polysaccharide chain, which increases the monotony. Variety can be introduced in the joining process, however, in a way that has no analogy in the English language. For example, one extremely prominent sugar, glucose (sometimes called dextrose commercially) forms polysaccharides of great importance. If the glucose units are all linked together in one way, we obtain starch, which we eat in bread, potatoes, and other foods. Should the glucose units be linked together in the opposite way, the result is cellulose, the principal ingredient of paper and cotton. The chemical difference is slight, but our bodies care. We can digest bread but not cotton.

Starch and cellulose illustrate two common biological uses of carbohydrates, as a food reserve and a structural material, respectively. Variety is not needed for these purposes. The situation changes, however, when we consider the remaining classes of molecules important to life, proteins and nucleic acids.

Proteins: They Do the Work

Diversity abounds in the proteins, the class of large molecules whose construction most resembles language. Twenty different individual units, called amino acids, are used by living organisms to construct proteins. The amino acids occur in a linear, variable, nonrepetitive order, as words do in the English language. Amino acids have great importance in human nutrition. We have lost the ability to make almost half of the necessary set for ourselves, and must therefore acquire them in the foods we eat. Bacteria, despite their tiny size, are much more versatile. Given a single carbon-containing substance such as glucose and inorganic sources for the other necessary elements, they will happily manufacture the entire set of twenty amino acids and every other organic compound that they need.

We do not understand why all of the life we know on earth has selected this particular set of amino acids to build proteins, out of the many thousands known to chemistry. The two simplest of the set, glycine and alanine, should be remembered, as they will be important at a later point in our narrative.

Nature has been quite selective not only in its choice of this exclusive set of twenty amino acids to do its work, but in an additional way as well. All but the simplest one, glycine, occur in two mirror-image forms. Mirror-image forms of a compound contain the identical set of atoms connected in the same way, yet they are not the same substance. They relate in the way that a right-handed glove does to a left-handed glove. Not all organic structures, but a great many of them, can occur as two such forms. To find an analogy, we need only look at our own handwriting. Certain letters when printed, c, a, g, and e, for example, will differ from their mirror-image forms, while others, such as t, o, i, and l, will be identical.

Sets of mirror-image forms have arbitrarily been termed right-handed and left-handed, using prefixes D- and L-, originally from the terms "dextro-" and "levo-," signifying right and left. Biology employs only left-handed amino acids in proteins. On the other hand, the sugars in living organisms are predominantly right-handed. The reason for this choice is again a mystery, and a subject of continuing dispute. The difference between the forms, slight in physical and chemical terms, is vital to us. If we were fed amino acids and sugars of the wrong mirror-image forms, we would starve.

Proteins, like polysaccharides, are used for construction purposes in biology. We encounter them as hair, leather, silk, and wool. This is not their most vital role. A subclass of proteins, called enzymes, has exquisite importance. The enzymes serve as biological catalysts, speeding up the chemical reactions vital to life. In short, they do the work and make things happen in the cell.

Nucleic Acids: They Hold the Plans

No less vital a role is played by the nucleic acids, the last important class of biological molecules that we will consider. Nucleic acids contain the genetic information of a cell, the actual instructions needed for the cell to do its work. The "language" of nucleic acids occurs in two closely related "dialects," DNA and RNA. The ultimate storehouse of information, the substance of our genes, is DNA. The orders contained within the DNA of a cell will determine whether it is a bacterium or will develop into a tree or a human being.

In the design of DNA, biology has adopted the same basic plan used for proteins and polysaccharides. Once again, a giant molecule is put together by linking many subunits in a row, with water split out to make each connection. The subunit used to construct a nucleic acid is called a nucleotide. Nucleotides are more complicated than amino acids and sugars, however. Each nucleotide is itself assembled from three smaller parts (or sub-subunits), called base, sugar, and phosphate. These parts are united in a very precise way (only one of dozens of possibilities is selected), and two molecules of water are split out in the process. When nucleotides combine to form a nucleic acid, sugars link to phosphates to form a long chain containing the two in alternation. The bases dangle from this chain, like so many charms on a necklace.

These bases perform the actual function of information storage. Four different ones are used in DNA, and the order in which they occur along the chain carries the information, as words do in a sentence or numbers within a computer. The physical difference between you and me, however great it may be now, was once encoded only in the order of bases along the DNA chains of two fertilized eggs.

We have not yet described the full complexity of DNA. In living cells, two chains or strands of DNA are wrapped around each other to form a structure called a double helix. Within the helix, each base on one chain encounters a partner on the other one and forms weak chemical bonds with it. Each base requires a specific partner for this purpose; they do not unite at random. A union of two appropriate bases within DNA is termed a base pair. Because of this requirement of each base in one chain of a double helix for a particular mate, the order of bases along one chain determines the order on the other one. The same information is carried in different form along each strand. The rules governing the base pairs and the structure of DNA were deduced by James Watson and Francis Crick at Cambridge University in 1953. Their contribution is considered a landmark of modern science, a cornerstone of the structure of molecular biology.

The other nucleic acid, RNA, serves not to store information but to ensure that the orders listed in DNA are carried out. It plays a number of roles in doing this, as we shall see shortly, when we follow our bacterium on an adventure. The nucleotide units used to construct RNA differ only slightly from those in DNA. RNA, however, occurs in most cases in single-stranded form, not as a double helix.

The function of RNA in living cells today is, then, to act as an intermediate in information transfer from DNA to protein. This may not always have been the case. Later we shall consider speculations that RNA evolved before DNA and served as the information storehouse of cells for a time.

We have described the major "languages" of biology, the large molecules used to build a cell. We must now depart from the language analogy. A book contains information, but it does not apply it. Bacteria, like other living creatures, do things, and alter themselves as they go about their business. To appreciate these aspects of bacterial life, we will follow the adventures of one bacterium for a time, viewing it with the aid of COSMEL.

Bacterial Days

As we enter the scene, our bacterium has located a supply of glucose and is grazing on it. It does not use a mouth for this purpose, as it has none. Glucose molecules pass through a rigid netlike outer cell wall and approach the membrane which lies just within it. The cell wall gives the bacterium its characteristic shape, and protects it mechanically. Bacteria placed in fresh water would swell up and burst, if not supported by their walls.

The lipid membrane protects the interior of the bacterium from foreign substances. It would not serve the purposes of the bacterium, however, if no passage through it existed. A number of gates, made of protein, control the movement of material in and out. Individual glucose molecules encounter the membrane and are welcomed at the appropriate portals. They enter readily, like roaches into a "roach hotel." Once inside, they do not emerge again. As they enter, they are tagged by attachment of a phosphate. Thus marked, they are retained within the cell.

What fate awaits them? They are devoured, digested, used as food to supply needed energy. They meet this end everywhere in living systems, in bacteria and in us. Life, like an appliance or an automobile, must have energy to run. A rock may persist, undisturbed by its environment, for millions of years. Unlike the rock, the chemicals in us are far from their state of greatest stability, which is called equilibrium. They resemble more a series of balls kept in the air by the constant activity of a juggler. A more or less continual supply of energy is needed to maintain this activity.

To provide a more relevant example, let us suppose that our bacterium wishes to build a new protein. Amino acids must be connected together, releasing water. Water is abundant both within a bacterium and in its environment. The production of additional water would be as welcome as coals in Newcastle. The favored process, the one that moves toward equilibrium, would be the reverse one: the disassembly of existing bacterial proteins with the consumption of water molecules. Our bacterium, however, wishes to construct additional proteins, rather than to see itself destroyed. For this purpose, it requires energy. A principle called the first law of thermodynamics stipulates that energy cannot be created (or destroyed) but simply converted from one form to another, within certain limitations. Our creature requires an energy source.

Glucose, and virtually every other organic molecule, represents a supply of chemical energy. When combined with the oxygen available in most environments on earth, it reacts to form carbon dioxide and water, releasing the stored energy. This reaction occurs quite rapidly at higher temperatures, as we can observe by putting a flame to sugar or a match to paper. At normal temperatures, such reactions occur too slowly to matter—which is fortunate for us, as we would otherwise combust and decompose on exposure to air.

Let us return to our glucose-phosphate combination within the bacterium. This molecule moves at random and in doing so encounters a series of enzymes which dismantle it, step by step. Oxygen is used in this sequence, and ultimately carbon dioxide and water are produced. When glucose is burned in a flame, the energy stored within it is released as heat. In the process controlled by enzymes, however, some of this energy is trapped, stored in another molecule called ATP. The energy within ATP is released when needed for construction of a protein or for other purposes of the cell.

What property permits each enzyme molecule to carry out its own special function? It is primarily the exact three-dimensional shape, governed by the precise order in which its component amino acids are connected. If we traveled about our model bacterium we would see a host of enzymes and other proteins of diverse shapes and sizes performing the work of the cell. Proteins would be building various macromolecules, repairing others, transporting materials about, and securing the energy supply.

At the start of the events we described, our bacterium had an ample glucose supply. Glucose is not a natural raw material on this planet like sand and water. Where did it come from?

The ultimate source of most energy for life on earth is the sun. Solar energy is captured in a process called photosynthesis. Various plants, ranging in size from sequoia trees to microscopic prokaryotes called blue-green algae, carry out this function. In the most important form of photosynthesis, carbon dioxide, water, and visible light from the sun are used to produce carbohydrates and release oxygen, thus reversing the process within the bacterium. Indirectly, the bacterium was fueled by the light from the sun.

Bacteria can exist using sources of energy other than glucose. A variety of organic molecules will serve the purpose, and some species have learned to release energy by combining inorganic chemicals such as sulfur or iron with oxygen. In principle, any suitable chemical reaction that releases energy might be adapted to sustain life. In our tale of events, however, we wish to challenge our bacterium in a particular way. Suddenly the glucose supply runs out and it encounters a less familiar carbohydrate called lactose. The lactose molecule normally occurs in milk. It contains two sugar units, glucose and another one, connected together in an uncommon way. As our bacterium moves about in this lactose supply, some molecules find their way through a gate and enter the cell. They cannot be digested, however. The two sugars must be disconnected first, and no enzyme for this purpose exists within the cell.

To dramatize this crisis, we will first describe it as a fable and then, more accurately, in molecular terms. The fable begins in a room marked "Bacterial Control Center." Inside, the committee of elves that control this complicated entity are wringing their hands.

"What a mess," says one. "We've run out of the good fuel, energy supplies are low, and they're shipping in this strange stuff instead, but nobody knows how to handle it. What'll we do?"

"We'd better consult the emergency part of the operating manual," replies another elf. They rush to a large dusty room filled with locked files. They fumble with a set of keys, and upon opening the files rummage around in panic. After a time, a shout of triumph is heard. One elf pulls out a folder containing an exact description of the new fuel and a set of blueprints for the construction of a machine that can use it.

"Rush these plans down to the shop," orders the chief elf. "I hope they have the parts to put this gizmo together." Fortunately, the standard parts needed to construct the unit are in good supply. Before long, the new machine is chugging away, burning lactose for fuel. Eventually, however, the bacterium moves out of the milk supply and enters the sugar stream again. At this point, the lactose-eating machine is dismantled, and the parts are used for other purposes. The plans are carefully returned to their place in the file and locked up again.

In the real world of a bacterium, the rough equivalent of these events takes place. The file is DNA, the plans are carried by an RNA molecule, the shop is the ribosome, the spare parts are amino acids, prepared in a special way for assembly into protein, and the new machine is an enzyme that can attack lactose.

When DNA exists in double helical form, its information is stored away, as in a closed filing cabinet. To open the file, the strands must be separated in the area containing the information desired. This process, aided by proteins, occurs continually during the normal operation of a cell. A copy is made of the desired stretch of information by assembling an RNA molecule of moderate length that matches the sequence of bases of one of the chains of DNA. Normally, the term "gene" is used for a length of DNA containing sufficient information to build one protein.

The RNA copy containing information from the DNA file is called messenger RNA. When the messenger has been constructed and has departed, the DNA helix closes up again. The messenger takes its message to a ribosome. This structure acts as an assembly line for the construction of proteins. The amino acids needed are brought to the ribosomes by various short molecules of RNA, of a class called transfer RNA. Each transfer RNA molecule specializes in the transport of a single kind of amino acid. The ribosomes themselves are intricate devices, each made of more than fifty protein and RNA molecules (the latter called ribosomal RNA) assembled into a specific three-dimensional array. Within the ribosome, the information present in the order of bases in the messenger RNA is used to direct the construction of a protein containing a particular order of amino acids. This process of information conversion from the language of nucleic acids to that of proteins is given an appropriate name, translation. The rules that govern this conversion, called the genetic code, are virtually universal in all known organisms, from bacteria to human beings.

The overall flow of information described above, from DNA to RNA to protein, has a central role in the processes of life on earth. It is described as the Central Dogma of molecular biology, first pronounced by Francis Crick. Sometimes, this rule is shortened to the phrase "DNA makes RNA makes protein."

We have described the normal operation of a cell, but the crisis concerning lactose differed in some respects. The lactose file was not merely closed, but locked as well. The lock, a particular protein molecule, sat on the DNA helix in the vicinity of the gene for the enzyme that could attack lactose, denying access to proteins that would open up the helix. Fortunately, a key could open this lock, and the appropriate one was the lactose molecule itself. When some of the lactose molecules entered the cell, one found its way to the DNA and combined with the protein "lock," removing it from the helix. The way was open for production of the needed enzyme.

At a later point, when all available lactose had been digested by the new enzyme, this key was attacked and destroyed as well. The protein lock was free to return to its position on DNA, shutting down the gene for the enzyme that digests lactose. The bacterium returned to its normal mode of operation.

In the above events, DNA served only as an information storehouse, but it has one additional vital function in the life cycle of a bacterium. Let us suppose that our specimen has prospered on its diet of glucose and lactose and enlarged its size considerably. At a certain point, it will solve its overweight problem by splitting into two bacteria. In preparation for this event, a complete copy of its DNA double helix must be made, so that each member of the next generation has a full set of instructions for life. At least twenty protein "midwives" as well as bits of RNA assist in this copying procedure. The entire DNA double helix, which in a bacterium may contain 4 million nucleotide units in each chain, is pulled apart in stages. A new partner chain is assembled to match each of the original ones. Once this process is completed, the remainder of the assets of the cell can be divided, and additional portions of membrane and wall built, to make the separation final.

The above narrative can only hint at the complications that fill the lives of bacteria. Volumes have been written on the subject. For our purposes, we need not learn these details, but we must remember that they are complex creatures, despite their size. We will now turn our attention to larger forms of life.

The Biochemical Unity of Life on Earth

Let us return from the lower levels of COSMEL and consider familiar plants and animals with a fresh eye. At first glance, their diversity is the feature that overwhelms us. Bees, trees, and chimpanzees appear to have little in common. Diversity is also striking at the cellular level, even within a single organism. Nerve cells, fat cells, and muscle cells all look and behave very differently. All of these eukaryotic cells appear quite complex when compared to a bacterium. If each cell type and every creature had its own distinct set of chemicals and basic cellular organization, the study of biochemistry would be endless. Fortunately, this is not the case. Underlying all these variations is a basic biochemical similarity.

In every known organism, heredity is preserved in nucleic acids, proteins are made in ribosomes, the same set of amino acids is used to construct proteins, energy is stored in ATP, and an almost identical genetic code is employed. Many other features are the same. In the same way that a child's construction kit can be used to build a toy house, bridge, or Ferris wheel, the same parts in the biochemical kit are used to construct the various forms of life known to us.

Once this concept is understood, then we can appreciate the variations in the common theme that do occur at the biochemical level, and use them as clues in tracing the path of evolution. One striking difference between eukaryotes and prokaryotes involves the way that genes are organized within DNA. If, for example, we imagine that a bacterial gene reads something like this (in the language of DNA, of course): "Here are the plans for the construction of a protein to digest lactose," then an equivalent gene in a higher organism might read this way: "Here are the plans gibble gabble for the construction of hubba hubba a protein to digest lactose." The story is interrupted by "commercial breaks," irrelevant messages that biochemists call introns. Their purpose is unclear. If eukaryotes arose in evolution from prokaryotes, then we must deduce why this evolution would involve the insertion of extraneous matter into perfectly good messages.

It is clear, however, that these interruptions never reach the ribosome. If they did, they would be treated as if they were an intended part of the message and a faulty protein would be produced. Rather, they are edited out in a splicing process at the RNA level.

Other nuances exist, but not in sufficient number to disrupt the basic message of the biochemical unity of life on earth. It is surprising to find this unity. We might have expected to see competition between biochemical systems just as we have competitions among species. If such competition existed earlier in the history of life, it was settled in favor of the system we know. We can infer that at some point in evolution an organism emerged bearing all the features common to life today. This organism prevailed, and inherited the planet. We have all descended from it. This hypothetical creature is usually called the last common ancestor.

Now that we have appreciated something of the construction of life, we are ready to explore the historical record and move backward in time in search of its origin.

three

The Testimony of the Earth

We have seen the complexity, the intricacy of organization that is abundant in even the simplest bacterium today. To learn how life came to be that way, we must turn to the record of the past. As we shall see later, Creationists argue that we cannot learn about the origin of life by scientific inquiry. No eyewitness was present; no human testimony exists to guide us.

One witness to these events remains with us, however: the earth itself. Our planet carries the record of its past in its sediments, mountains, and valleys, just as our own bodies do in their scars and wrinkles. Also preserved within the earth are fossils, impressions and replicas in stone of the life forms that once inhabited it. To create a meaningful history from them, we need to place them in sequence, ideally with a firm date attached to each. To this purpose, scientists have struggled for centuries to determine the age of the earth, and the rocks and fossils within it. We will review this historical achievement, as it illustrates nicely the way in which science arrives at a firm conclusion by a fitful series of approximations that gain increasing accuracy. The tale also serves to contrast scientific and mythological approaches to the same question. To begin, we might ask how it is possible to learn the age of anything.

We can specify the age of an object by stating how much time has passed from the moment when it was created until the present. We need some ruler, however, some measuring device on which we can indicate the point when a past event occurred. The device must be a process which we can follow, and which takes place at a constant rate in time. For events in recorded history, the passage of day and night has served admirably. The seasons have provided an additional measure, allowing blocks of days to be grouped together into years. When humans developed the ability to write and count, a calendar could be kept. Events were identified with a given date and year, and we could calculate how much time had passed since they took place.

Modern archeologists have discovered records in cuneiform writing in Iraq that were made about 3000 B.C. and Egyptian hieroglyphic writings that are almost as old. Until the nineteenth century, however, the Old Testament was the oldest continuous record of events known to Western civilization. According to the Bible, the creation of the earth, life, and man all took place within the same week. The purpose of the earth was to provide a home for mankind. There seemed no reason for it to have a lengthy history before the time that humans appeared.

The Bible provided no exact date for the time of creation, but did record the passing of each generation and the age at death of prominent figures. The age of the earth could be estimated from this information, and appeared to be a few thousand years. To put matters on a firmer basis, theologians in the Middle Ages presumed that the six days used for the actual work of creation represented as well a period of six millennia allotted for the duration of all of human history. After this period, the Second Coming of Christ would bring the final millennium, His Kingdom on Earth. His arrival often seemed imminent, so the assumption was made that the earth was about six thousand years old. In Shakespeare's *As You Like It* (Act IV, Scene 1), Rosalind says, "The poor world is almost six thousand years old . . . " reflecting common understanding.

As modern society developed, greater precision was desired, even in religious matters. Archbishop James Ussher of Trinity College in Dublin consulted ancient Hebrew texts, the Hebrew calendar, and the Bible and concluded in 1650 that God had made heaven and earth on the evening of Saturday, October 22, 4004 B.C. By modern scientific standards we would say that he had used too many significant figures and should have rounded off his estimate. John Lightfoot of Cambridge University, a contemporary of Ussher, simplified his estimate, stating that creation had taken place in September 3928 B.C. Ussher's data prevailed, however. It was listed as a marginal note in an English Bible in 1701, and was republished thereafter. As we shall see, certain religious groups still maintain on the basis of biblical authority that the earth is only a few thousand years old. For an alternate view, we must turn to a very different discipline: science.

A Time for Science

Scientists do not presume that the earth is no older than mankind. Written history serves only to establish a minimum age for the planet. Other measures are needed to determine how long the earth may have been here before our civilization took notice of it. Periodic phenomena are the easiest to use for this purpose. Tree rings are readily counted, for example. We have learned from experience that occasionally a tree may skip growth in a year or form two rings in a season. But for the most part, the rings accurately reflect the age of the tree. The oldest trees in existence, in California, testify to a history extending back more than 4,000 years. No older life has been discovered, so we must turn to geological events to penetrate further back. Certain glacial lakes deposit dark bands of clay in the winter and light ones in the summer. Based on these deposits, called varved clays, some lakes in northern Europe can be assigned an age of 8,700 years. The lakes thus predate biblical chronology. Regular annual phenomena do not take us further back, but other geological evidence indicates that the earth is much, much older than the above figures.

In the eighteenth and nineteenth centuries some geologists, examining the appearance of the earth without any presumptions derived from mythology, inferred that it looked very old. The Scotsman James Hutton observed the slowness of processes such as rock weathering and sedimentation. He concluded in his book *Theory of the Earth* (1795) that our planet showed "no vestige of a beginning—no prospect of an end." His efforts were expanded by Sir Charles Lyell, a geologist who greatly influenced the thinking of Charles Darwin. From their work, and that of others, a point of view arose that placed the origin of the earth, and life, many millions of years in the past.

The accumulation of sediments served as an index of time in these estimates. The flow of water eroded rocks, and the soil that resulted was moved downstream by rivers. When a river reached a broad flat area, it slowed down and deposited its debris as a sediment. Such processes were observed even during recorded history. The Greek historian Herodotus, noting the annual deposit left by the Nile, estimated that the river had required many thousands of years to build up its delta. In 1854 a statue of Ramses II, dated from 1200 B.C., was found buried under 9 feet of river mud. An estimate was made that 3.5 inches had been deposited per century. In the Grand Canyon, about a 1-mile (1.6 kilometer) thickness of sedimentary rocks lies exposed, above the ground. Using the above relation, we might guess that 2 million years were needed to deposit that much sediment. The estimate represents a minimum, however, as sediments may be compressed when buried at great depth, and occupy less space.

No more than about 1.5 miles (2.5 kilometers) of sedimentary rock is exposed at any single site on this planet. Much greater thicknesses have been inferred by geologists, however. Sediments are not uniform, but show layers of irregular thickness when viewed in cross section, in a cut made by a river. These layers reflect changing geological conditions which led to the deposition of different kinds of mud at different times. Often, when the sequence of layers in different locations is compared, similarities are observed. The same type of event had occurred at the same time in both places. In any one locality, only a certain number of layers lie exposed. By correlating their observations from many sites, however, geologists can reconstruct a much greater sequence of layers than exist at any single location. This sequence is called the "geological column."

The reasoning involved in this process can be illustrated by an analogy using letters in a line. Suppose that many copies of this sentence we are reading now were printed, shredded at random, and scattered about the room. By gathering and examining the fragments, we could reconstruct the sentence. One fragment, for example, might read "many copies of th," while another would have "pies of this senten." By combining them we would assemble the sequence "many copies of this senten." Examination of additional fragments would afford the full message.

From the overlap of different incomplete exposed sediments, geologists have calculated a total sedimentary geological column of perhaps 75 miles (120 kilometers). In estimating the age of the earth, of course, not only deposition times are considered. Such sediments must have been uplifted from river- or seabeds by geological forces, then exposed by river erosion or other events. By estimating or guessing the rates of such processes, geologists in the nineteenth century arrived at figures of several hundred million years for the age of the earth.

A very different approach to this question was taken by other scientists, one that depended on the heat within the bosom of Mother Earth rather than the wrinkles on her face. Men had long realized that the center of the earth is hot. The temperature goes up as we go down in a mine shaft. Hot springs and lava eruptions also testify to greater heat deep inside the earth. Thinkers in the eighteenth and nineteenth centuries assumed that the earth had been formed in molten condition and gradually cooled down. The crust, exposed to outer space had solidified, but the interior was still in molten form. By considering the present condition of the earth and the measured rate of cooling of solids, an estimate of the age of the planet was obtained.

Isaac Newton had performed a calculation of this type, estimating that a red-hot iron sphere of the size of the earth would cool down in 50,000 years. He rejected this answer because of his religious convictions, assuming that some error had been made. He commented, "I should be glad that the true ratio was investigated by experiments."

The eighteenth-century French naturalist Georges Louis Le-
clerc, Comte de Buffon, was not restrained in this way, as he felt
that the biblical "days" of Creation represented longer periods.
Buffon performed calculations and experiments on the cooling
rates of spheres. After taking into account all factors that he felt
relevant, Buffon concluded that the earth was 74,832 years old, a
figure not very far from the rough estimate of Newton. By further
extrapolation, Buffon determined that after a further 93,291 years
had passed, the earth would become too cold to support life. Like
Archbishop Ussher, Buffon was too taken with his own calculations. In listing his numbers to the exact year, he suggested a far
greater precision than was warranted by his methods. He had
underestimated, for example, the amount of heat delivered by
the sun to the earth.

Improved calculations were made in the next century by the
famous British physicist and inventor William Thomson, who in
1892 became Lord Kelvin. Kelvin had made fundamental contributions to the mathematical theory of heat flow (a scientific temperature scale is named for him) and to other areas of physics.
After the publication of *The Origin of Species*, he was attracted to
the controversy concerning Darwin's theory of evolution.

As we have noted, geologists and advocates of Darwin's theory
had assumed a history of hundreds of millions of years for the
earth, which allowed ample time for the slow processes of evolution. In a series of papers published from 1862 to the last years of
the century, Kelvin calculated that the earth had a lesser age,
using data derived from cooling rates. In 1862 his estimate had
been 100 to 200 million years, but by 1897 his "irrefutable" calculations had shortened the age of the planet to 10 to 20 million
years. Other workers produced even shorter spans for earth's
history.

Darwin was aware of the difficulties that these physical limitations raised for his theory, but remained cautious. He wrote in
the final revision of his book: "Many philosophers are not yet
willing to admit that we know enough of the constitution of the
universe and of the interior of our globe to speculate with safety
on its past duration." His caution was justified. Little more than a
decade after his death, the discovery of radioactivity by Henri

Becquerel in 1896 changed the picture entirely. Subsequently, it was recognized that the release of heat by radioactive minerals within the earth was quite sufficient to compensate for its loss of heat to space and to maintain the planet's temperature. The calculations of Kelvin were irrelevant, and the geological view was closer to the truth. An observer of the controversy, T. C. Chamberlin, commented at that time: "The fantastic impressiveness of rigorous mathematical analysis, with its atmosphere of precision and elegance, should not blind us to the defects of the premises that condition the whole process."

Not only did the discovery of radioactivity demolish the basis of Kelvin's calculations concerning the earth's age, but it also provided a much better method for this purpose, one capable of recording ages much greater than those considered before. Atoms of hydrogen, carbon, and the other basic elements that make up the universe can occur in alternative forms called isotopes. Certain isotopes are unstable, however. Radioactivity involves the decomposition of unstable isotopes into a variety of products.

Thus, one isotope of potassium present in minerals decays slowly to yield calcium and the gas argon. About 1.3 billion years are required for half of any quantity of this unstable substance to decay. The products are retained in the rock, alongside the remaining potassium. By measuring the amounts of these materials in a rock, geologists can calculate the time that has elapsed since it first solidified.

Often a rock, or a series of related rocks, will contain more than one unstable isotope. The results obtained from one decomposition mode can then be compared to the others, and to the position of the rock in the geological column. These techniques have been applied by many scientists over the course of this century, and consistent dates obtained for those minerals of importance. Ultimately, a way was found to apply these methods to the age of the earth itself.

An Indelicate Question

"It is perhaps a little indelicate to ask our mother Earth her age, but science acknowledges no shame and from time to time has boldly attempted to wrest from her a secret which is proverbially well guarded." So wrote Arthur Holmes (1890–1965) in *The Age of the Earth* (1913).

The determination of the age of our planet by radioactive dating was not achieved easily; a long trial-and-error period of development was needed. As this process took place, improved estimates of the age of the earth were made continually, and it grew older and older in the eyes of scientists. The sentence quoted above was written by a noted geologist at a time when both he and the dating methods were young. The man and the technique matured together.

The earliest determinations gave rock ages that varied from 400 million to 2 billion years. The relative ages estimated in the nineteenth century were correct, but their magnitudes were too small by a factor of 10. The first radioactive dates also suffered from mathematical errors, and were 20 percent too large. By 1941 the oldest rock sample known had been assigned an age of 2.6 billion years. There were still a number of uncertainties, and a review article from that time stated that the earth "appears to have an age close to 2 billion years."

The oldest known rocks may of course be younger than the planet itself. In 1946 Arthur Holmes and F. G. Houtermans reported an indirect method by which the age of the earth itself could be estimated from radioactive data obtained from younger geological samples. Their estimate of about 3 billion years was accepted until 1953, when errors were discovered and a new age of 4.5 billion years was proposed. This figure is the accepted one today. The oldest known rock formation on earth, at Isua in southwestern Greenland, is younger, "only" 3.8 billion years old.

The indirect age deduced for the earth has been reinforced by studies of extraterrestrial bodies. Meteorite ages range up to 4.5 billion years. The oldest moon rocks have been dated at 4.6 billion years. As most theories concerning the formation of our solar system postulate that the various bodies in it were formed at about the same time, these findings increase our confidence that the indelicate question concerning Mother Earth has been answered correctly.

This confidence is important, since the final answer is so remarkable. It is far easier to comprehend a biblical history of

6,000 years, which amounts to about 80 human lifetimes. The geological age, however, computes to over 60 million human lifetimes. If the earth's age were made equivalent to one year, one lifetime would pass in the time needed for us to blink our eyes twice, as quickly as possible.

Natural Selection

Using modern dating techniques, geologists can assign ages to fossils. The predominant species at each time in history can then be identified. The results are quite surprising. While life itself has existed on earth throughout most of its history, creatures containing more than one cell are only represented in fossils formed within the last 800 million years. Worms, jellyfish, and other organisms made only of soft parts came first. They were followed by fish, land plants, amphibians, trees, reptiles, insects, birds, and mammals, roughly in that order. Some creatures, such as the dinosaurs, arose only to vanish again. The story of the evolution of higher life forms has been told and retold, and need not be repeated here. More important to our narrative is the mechanism responsible for this progressive appearance of life forms: natural selection.

Almost all scientists hold today that the more complex life forms on earth were derived from simpler ones, as described in the theory of evolution. The best understood of the mechanisms that drive this process is the one called natural selection.

The actual details of the changes remain a subject of controversy. The transformations may have taken place gradually, as traditional followers of Darwin believe, or more rapidly, according to a theory of "punctuated equilibria." Additional mechanisms beyond natural selection may exist, a possibility not excluded by Darwin. We will certainly learn much more about this topic in the near future, because of rapid advances in our knowledge of the cellular roles of DNA.

This substance, as we have learned, is the hereditary material of living organisms. In replication, a copy of DNA is made for transmission to progeny. Copying mistakes can occur in this procedure, giving rise to mutations, changes in the genetic mes-

sage. Geneticists have learned much about the mutational processes that produce changes equivalent to the alteration of a single word in a sentence. Recent discoveries have indicated that
much larger blocks of information also move about in our genetic material by natural mechanisms. These mobile segments of
DNA have been called "jumping genes." Much more familiar to
us, of course, are the variations in living creatures that are produced by sexual reproduction.

All of these mechanisms introduce variability into populations
of living organisms. Many of the variants that are produced, particularly those created by random mutations, are not necessarily
better than their predecessors. If you doubt this idea, try replac-
ing a word of this very sentence with one chosen at random from
the dictionary! In most cases, the fate of the unfortunate products
of this process is extinction. Occasionally, however, a mutant
may survive and inherit the future. We can illustrate this with an
example. Suppose that a strain of bacteria in a particular location
has been destroyed by a new antibiotic. A single individual of
the billions originally present has survived, however. It had acquired a genetic change that allowed it to resist the drug, while
the remainder perished. If other circumstances remained favorable, this single organism could reproduce and repopulate the
entire environment in a few days. The favorable gene would be
amplified, and become part of the genetic instructions of the
strain.

These events illustrate the process of natural selection, the
force that most scientists believe responsible for the process of
evolution. At a later point we will consider whether natural selection could also have served to create the first living creature.

The Era Of Microorganisms

A rich harvest of fossils describes the age when many-celled
creatures prevailed on earth. The last 600 million years, when
hard parts such as shells and bones were available for making
fossils, are particularly well documented. A much scantier record
remains of a longer era lasting at least 2.5 billion years, when life
was represented only by one-celled organisms. Until a few dec-

ades ago, in fact, there was some question as to whether any life at all existed on earth during that time.

This scarcity of data can readily be understood. Rocks can perish, by weathering or remelting. Recent ones are abundant, but the older ones are in shorter and shorter supply as their age increases. Finally, with the 3.8-billion-year-old Isua rocks of southwestern Greenland, the record vanishes entirely. Nothing remains to tell us directly of earlier days on this planet.

Another problem lies in the locating and identifying of fossils of microorganisms. Once uncovered, dinosaur bones leave little doubt about their identity. Microfossils, however, are less readily located. Their fossil nature is often ambiguous and tells us little more than the size and shape of the cell. Despite these difficulties, geologists have worked patiently at about forty locations to build up a picture of a long era of microorganisms, which lasted from 3.5 to 0.9 billion years ago.

According to this record, eukaryotic cells first appeared about 1.2 billion to 1.4 billion years ago. This date could be revised at any time, of course, if older eukaryotic fossils of a convincing nature should turn up. Claims have been made and disputed, but most workers in the field seem satisfied with the above range.

Remnants of prokaryotic forms that resemble modern bacteria and blue-green algae (the latter are also called cyanobacteria) go back to much earlier times, more than 3.5 billion years. This record is continuous and well documented to about 2.2 billion years ago and then extends intermittently backward to the oldest known fossils, which are located in western Australia and South Africa. For a long period of over 2 billion years, about half the age of the earth, only prokaryotes represented life on this planet. While the direct fossil imprints left by these ancient creatures are microscopic, other remnants of their existence are of visible size. At the Australian site, called North Pole for its remoteness (but not for its climate), a dome-shaped structure about a foot high can be seen, embedded in a weathered outcropping of rock. Such objects, made of hundreds of wafer-thin layers of rock, have been likened to cabbages, water biscuits, and fossilized baklavas (a type of flaky pastry produced in the Middle East). We can recognize them as products of life because their counterparts exist today. These structures, called stromatolites, are found in shallow water in restricted locations, such as the coast of Australia some miles from the North Pole site. They are produced when colonies of microorganisms, usually blue-green algae, grow in layers that accumulate pebbles and debris. A new layer of blue-green algae will then form above the debris, and the cycle will repeat itself. In the ancient fossils, the tenants no longer remain, but the houses built by them persist.

Modern blue-green algae prepare their own food by photosynthesis, using carbon dioxide from the air and solar energy. They do not make use of organic compounds found in their environment, as did the bacteria that we discussed in the last chapter. If the creatures that built the ancient stromatolites resembled modern blue-green algae, then the process of photosynthesis is a very ancient one. This conclusion is supported by other evidence, derived from the ratios of carbon isotopes in very old sediments.

Apart from the stromatolites, both the Australian and South African sites display direct evidence of cells that may have existed 3.5 billion years ago. The fossil imprint of a row of cells stuck together to form a bent filament, found in the Australian location, bears a striking resemblance to filaments on bacteria seen today. A South African fossil shows a series of spheres joined together, apparently in various stages of cell division. The studies at both sites have been extensive, and are well documented. They have led to a consensus that life was well established at more than one location by 1 billion years after the formation of the planet.

The caution shown, and the amount of documentation produced, have been necessary. Minerals also contain organized elements of an inorganic nature that at first glance may resemble biological fossils. Mistakes have been made. We can quote, for example, the evolutionist G. G. Simpson: "Eozoon, proudly named the 'dawn animal,' is now considered to be no animal at all, nor yet a plant or any form of life but a mere inorganic precipitate."

The eozoon was a phenomenon of the nineteenth century, but its spiritual descendants persist today. One recent emergence occurred in 1979 and concerned the Isua rocks. The discovery of evidence of life in very old rocks apparently aroused the desire in some workers to extend this process to the logical limit: to find fossils in the oldest rocks now known on earth. Unfortunately, the Isua rocks are ill-suited to this purpose, as they were heated considerably several times during their history. Such treatment usually destroys fossils. In spite of this, two separate reports claimed the discovery of strong evidence for life at Isua.

The American Chemical Society met in Washington, D.C., in the summer of 1979, and its news magazine, *Chemical and Engineering News*, reported the exciting tidings. A headline proclaimed: "Evidence of Life Found in Oldest-Known Rocks." A group of scientists from various universities had presented a paper at the meeting. Their spokesman at a press conference was Cyril Ponnamperuma, of the University of Maryland. The report in the magazine told of the isolation of hydrocarbons from the Isua rocks, with evidence from the carbon isotope ratio. The ratio suggested that the compounds had been prepared by photosynthesis, and hence by organisms. The report mentioned that actual fossils, which would provide more convincing evidence of life, had not been found thus far.

This deficiency was remedied by others. An article appeared at about the same time in the prestigious British scientific journal *Nature*, written by a German geologist, H. D. Pflug, and a colleague in France, H. Jaeschke-Boyer. They had found "cell-like inclusions" at Isua, which they identified as fossils of ancient microorganisms. Single cells, filaments of cells, and colonies were observed. They coined the name Isuasphaera for their find and commented: "There is little doubt that Isuasphaera is an organism." In particular, they were impressed with a sheath that surrounded their creature: "The exterior multilaminate sheath enveloping the Isuasphaera cell can only be understood as a product of biological activity." Vacuoles, hollow areas found in some living cells, were observed, as were buds that resembled those produced by yeast cells. The authors felt that their organism resembled yeast, but tempered their enthusiasm by noting that yeast is a eukaryote. They were reluctant to claim the eukaryotes had arisen so early in evolution and suggested that Isuasphaera had some intermediate status.

A certain absence of doubt and skepticism in these reports should serve as a warning flag for us. In cases where scientists are unwilling to play devil's advocate for themselves, others are usually quite willing to oblige. In this case the rude awakening came eighteen months later, in several articles also published in *Nature*. The presence of hydrocarbons at Isua was confirmed. If one analyzed further, however, not only hydrocarbons but also amino acids could be detected, including some rather perishable ones. In fact, the age of the entire chemical mixture could not have exceeded some thousands of years, let alone billions. The amino acid composition resembled that present in lichens growing on the surface of the rocks at the present time. The inference emerged that chemicals from the plants on the surface had penetrated into the interior of the rocks in the relatively recent past.

Isuasphaera suffered no kinder a fate, and was dispatched down the same path as Eozoon. The supposed fossils were examined by an international team which included scientists prominent in the studies of the Australian and South African fossils. They concluded that the Isua structures were demonstrably inorganic artifacts and did not constitute evidence for life. In a wonderful example of scientific understatement they described their paper, in its title, as a "cautionary note."

With the destruction of these claims concerning Isua, we are left with a state of incomplete knowledge. Prokaryote-like forms existed 3.5 billion years ago, but we do not know the time or the circumstances of their origin. At this point the fossil trail runs out.

The Rise of Oxygen

Little evidence of evolutionary progress is revealed by the shapes and sizes of microfossils derived from the era of microorganisms, despite the fact that the age lasted almost half the history of this planet. Important changes may have taken place within these ancient microorganisms during this period, even though their external appearance varied little. Again, the planet itself is the main source of evidence.

Earth alone, of the known worlds of the solar system, has a sizable fraction (20 percent) of oxygen in the air. As we have seen, oxygen is vital to the processes of all cells of higher organisms. It is used to combine with foodstuffs, producing carbon dioxide and water and releasing energy. Only some bacterial species are exempt from this requirement, as they can obtain their needed energy from reactions which do not require oxygen gas.

The rock record indicates that our atmosphere was not always as rich in oxygen as it is today. Striking changes in the nature of the minerals deposited in rocks took place roughly 2 billion years ago. In particular, the formation of characteristic banded iron formations occurred widely at that time but not thereafter. Perhaps 90 percent of the known iron-rich ores that constitute our current source of that metal were deposited then. These changes are believed to be the result of the first appearance of large amounts of oxygen in the atmosphere.

The source of the oxygen is a different question. Most scientists agree on the primary cause of this transformation: it was the release of oxygen in photosynthesis by organisms such as blue-green algae.

We have discussed the evidence that photosynthesis may have existed as long as 3.5 billion years ago. It is an open question whether the very first organisms prepared their food in this way or used organic compounds available in the environment. Whatever the answer to this, photosynthesis with release of oxygen started at some point in earth's history.

For a time, the liberated oxygen may have been consumed by a reserve of substances in the environment that can combine with it. When these ran out, oxygen accumulated in the air. This change may have poisoned many organisms, causing their extinction or driving them to sanctuary in specialized oxygen-free niches. Some such species survive to this day. One prominent group of this type, called the methanogens, is killed by oxygen and inhabits locations such as the mud at the bottom of the Black Sea and San Francisco Bay. Methanogens obtain energy not from oxidation, but from other chemical reactions. Their lifestyle, and certain chemical differences that make methanogens distinct from most other bacteria, have stimulated speculation that the methanogens have survived from earth's earliest days. At that time, the atmosphere was more congenial to them and provided the gases used by them to release energy.

The oxygen-rich atmosphere, while toxic to some species, was a boon to the organisms that adapted to it. They could obtain much more energy by combining organic compounds with oxygen than they could get from the earlier methods that they used. This advantage may have stimulated the development of eukaryotic cells. The new atmosphere also provided an additional beneficial effect. A series of complex reactions in the air led to the formation of some ozone gas, an alternative form of oxygen. Ozone absorbs a particular kind of solar radiation called ultraviolet light. This radiation is harmful to many of the chemicals present in living things. Before the development of the ozone screen, the land and upper layers of the sea may have been uninhabitable. Thus the colonization of land by living creatures may have become possible only after the introduction of oxygen into the atmosphere.

The Oldest Rocks

Many problems remain concerning the details and the timing of the conversion of earth's atmosphere to its present form. These matters are important, but are not as crucial to the origin of life as another question: What was the nature of earth's atmosphere prior to the oxygen release, when the first living cells appeared? For the earliest testimony, we must turn to the Isua rocks.

The heating process that they experienced may have destroyed any fossils that existed, but it did not obscure the basic geological messages that were present. These rocks are sediments, laid down at the bottom of the sea and made of particles formed by erosion of other rocks. These earlier rocks were volcanic cones, rather than continental material. For various reasons, geologists believe that continents formed at a later date. At the time of Isua, the earth was covered by a shallow sea, with most of the land mass, which was less extensive than now, consisting of volcanic material. The Isua rocks, and other ancient sediments, are otherwise unremarkable in their composition, containing many mineral substances which are familiar to us today.

This evidence is quite skimpy, but if we attempt to probe further back in time in earth's history, we have no concrete evidence to examine at all. In such circumstances, scientists fall back on other techniques, using the known laws of chemistry and physics to construct models. A model is considered successful if it starts from a set of plausible initial conditions and, using established laws, calculates that the present state of affairs would result from the initial one. We cannot be certain, of course, that a particular model is the best one, as some very different set of conditions not yet thought of may do the job in a better way. A model, no matter how vulnerable to change, is still better than no structure at all, so we will consider current scientific thinking about the formation of the solar system and the earth.

The Birth of Planet Earth

When I was young, I read that the solar system had been formed in a near encounter of two stars which caused enough matter to be drawn out of them to form the planets. This theory has fallen out of favor. The chances of such a near collision are slight, and more important, the mathematical models of such an event do not produce a planetary system with the characteristics of our own. The current paradigm, one accepted in its general form but not in all its details, is the nebular theory. According to this idea, the sun and planets were formed at the same time by the condensation of a cloud of interstellar gas and dust.

We can observe such clouds today elsewhere in the galaxy, some apparently in the process of giving birth to new stars. Their composition reflects that of the universe as a whole: mostly hydrogen and helium (a light, inert gas used in balloons), with small amounts of other elements. The process of star formation begins when an interstellar dust cloud starts to contract by gravitational attraction. The contracting cloud, called a solar nebula, takes on a spinning motion and a disc shape. Most of the material in it collects at the center, and heats up because of gravitational forces. After a certain temperature and mass have been reached, a nuclear reaction begins, converting hydrogen atoms to helium. When this additional and durable energy source switches on, the life of a star has begun.

According to the nebular theory, these processes led to the birth of our own sun, more than 4.5 billion years ago. Not all of the matter in the nebula ended up in the sun, however. Additional debris remained in orbit at various distances, and collected to form planets, satellites, meteorites, and comets. The chemical composition of each body was in part controlled by its distance from the sun, which determined the temperature of the nebula at that point. At the distance of the earth, heavier iron and lighter silicate minerals could exist in solid form, and these came together to create our planet.

Different theories have been put forward to describe the collection process. All of them contain a common feature: they must conclude with the planet in its present form. We have learned this condition from the study of shock waves produced by earthquakes, of the earth's magnetic field, and of other features. The interior of the planet is composed of several distinct zones. At the center exists a core largely made of solid and liquid iron. An intermediate zone called the mantle, composed of partly molten rock, lies over it. At the top there is a thin crust, a few miles thick, and above this familiar rock layer are the oceans and atmosphere.

The most direct theory of earth's formation suggests that this existed from the beginning. Iron particles aggregated first to form the core. When it had grown to a certain size, its gravitational attraction caused the less "sticky" silicate materials to layer on top of it. Another model suggests that rocks of various composition aggregated to form larger bodies, called planetesmals. As this collection process continued, the earth grew in size, requiring perhaps 100 million years to reach Martian dimensions. At some point, the heat released by gravitational forces, augmented by heat from radioactivity, caused the interior of the planet to melt. This permitted the heavier iron to sink to the center as the light silicate components floated to the top.

Whatever the mechanism of its creation, it was likely that the surface of the earth was in a turbulent state after it formed. The cratered appearance of the moon and other airless bodies of our solar system testify to a period of intense meteorite bombardment which ended perhaps 4 billion years ago. At one point during the development of the sun, an intense burst of emitted solar material and radiation, called the solar wind, is believed to have swept the solar system clean of small debris. It also stripped away any atmosphere that the earth may have inherited from the solar nebula. For a time, the earth may have existed in a cratered airless state, as the moon does today.

This situation could not endure. The changes taking place within the interior expressed themselves at the surface in the form of volcanic activity. A new atmosphere, the predecessor of our current one, was formed from gases released by volcanoes. The scarcity of elements such as the inert gas neon in our atmosphere, when compared to its much greater abundance in the sun, testifies to this internal source of the air surrounding the early earth.

A key question for the origin of life concerns the nature of this early atmosphere. A critical distinction lies between oxygen-rich environments (called oxidizing or oxidized) and hydrogen-rich ones (reducing or reduced). Oxygen and hydrogen are not likely to remain together for long, at least at earthly temperatures. Any spark, shock, or catalyst will enable them to react with one another, often violently, to produce water. Earth's atmosphere now has a strong oxidizing nature. In the universe generally, with its overwhelming hydrogen content, reduction reigns supreme. The nebula from which our solar system formed was also likely to be rich in hydrogen, as was earth's initial atmosphere. The large, mostly gaseous outer planets such as Jupiter remain that way today.

If we accept the volcanic origin theory for our present atmosphere, however, then the fate of the initial reducing one is well described by the phrase "gone with the wind." Life began when the later one was in place. Some clues about its composition can be obtained by sampling the gases released by volcanoes today. On this basis, most geologists believe that the earliest form of our present atmosphere contained nitrogen, carbon dioxide, and water, with small amounts of other substances. Hydrogen was present in amounts less than 1 percent, its accumulation limited by its escape into space. After sufficient water had been released, this moisture condensed to form rivers and seas. We then arrive at the world suggested by the Isua rocks, with an atmosphere above it that was neither oxidizing nor reducing but rather neutral, with perhaps a slight reducing character.

No great certainty surrounds this answer. A planetologist remarked at a recent conference that "the history of the early earth is among the most obscure and intractible problems that we have to face." This very ambiguity, so typical of pre-paradigm areas of science, is nonetheless the background that must be used for theories concerning the origin of life. Some current theories, including the best-known one, require an alternative environment, and this disagreement contributes to the confusion surrounding the origin-of-life question. We will explore this situation in the next chapter, when we take up the dominant paradigm in this area.

four

The Spark and the Soup

In 1952, Stanley Miller, a young graduate student at the University of Chicago, performed an experiment that had a profound effect on scientific thought concerning the origin of life. He exposed a mixture of reduced gases to an energy source, an electric spark, using an apparatus that he had designed in consultation with his research advisor, Professor Harold Urey. The reaction products included significant amounts of two amino acids that are among the twenty used by living cells to construct proteins. Upon their publication in 1953, the results were picked up by the media. *Time* reported that Miller and Urey "had simulated conditions on a primitive Earth and created out of its atmospheric gases several organic compounds that are close to proteins. . . . What they had done is to prove that complex organic compounds found in living matter can be formed. . . . If their apparatus had been as big as the ocean, and if it had worked for a million years instead of one week, it might have created something like the first living molecule."

The very circumstances of the reaction may have reinforced its impact on the public. Over the previous two decades, Hollywood had released a series of films in which dead matter was brought to life by the action of electricity. The original release, in 1931, was described in the following way by a historian: "The creation sequence was visually exciting, with electrical pyrotechnics setting a fine example for future film versions." The product of this transformation, however, was not amino acids but the Frankenstein monster, played by Boris Karloff.

In the case of the Miller-Urey experiment, the scientific community was impressed as well as the public. The work was cited repeatedly in the following years, incorporated into high school and university biology and geology texts, and featured in museum displays. It has become the classic, best-known experiment on the origin of life. Numerous variations have been performed, using a variety of energy sources, to create a formidable literature on the subject. One account that summarizes its impact comes from the chemist William Day:

> It was an experiment that broke the logjam. The simplicity of the experiment, the high yield of the products, and the specific biological compounds in limited number produced by the reaction was enough to show that the first step in the origin of life was not a chance event, but one that had been inevitable. . . . With the appropriate mixture of gases, any energy source that can cleave the chemical bonds will initiate a reaction resulting in the formation of life's building blocks.

The implications of the experiment extended to a wider range of problems than the origin of life on earth. To quote the astronomer Carl Sagan: "The Miller-Urey experiment is now recognized as the single most significant step in convincing many scientists that life is likely to be abundant in the cosmos." Any result that produces an impact of this magnitude deserves close attention, so we shall examine the actual data, and the interpretations placed on it, in some detail.

The Spark of Life

The equipment used by Miller had three essential components. The first was simply a flask of boiling water. As the vapors of water rose and passed out of the flask, they entered a compartment that contained two electrodes. Sufficient voltage was maintained between the electrodes to cause a spark discharge to jump across the gap that separated them. After the vapors had passed through this discharge, they entered a cooler area, where they condensed to form water droplets. These droplets then flowed back into the flask. The entire system was sealed off from the atmosphere and filled with a mixture of methane, ammonia, and hydrogen gases. The overall concept was simple; *Scientific American* has published an article for amateurs describing how they can construct their own Miller-Urey apparatus.

The original experiment was run for a week. As it progressed, the water in the flask took on first a red and then a yellow-brown color. After a week the spark was shut off and the contents of the flask were analyzed by various chemical methods. During the course of the experiment, the methane had been consumed and the carbon atoms originally present in it were now distributed among a number of organic substances. The predominant product was an insoluble material, made of a network of carbon and other atoms connected together in an extended, irregular manner. This substance coated the walls of the apparatus. Material of this type, called tar, resin, or polymer (a term meaning "many parts"), appears frequently in organic reactions. It constitutes a nuisance, particularly when the time comes to clean out the equipment.

Fifteen percent of the material had not been converted to tar and could be identified by chemical means. A list of the compounds present and their amounts was prepared. In any reaction of this type, the number of products identified depends primarily upon the patience and skill of the investigator. Instruments are now available that allow components to be identified at the level of a few parts per million, or even per billion. At such levels, many thousands of substances may be present in the reaction mixture. Harold Urey had been asked, in advance of Miller's experiment, what he expected would be produced, and answered "Beilstein." The name refers to a multivolume handbook which describes millions of organic compounds. According to Miller, "Urey's reply meant that the electric discharge would be expected to produce a little of everything."

If all the products had been present only in tiny amounts, little importance would have been placed on the experiment. The actual result differed in that a few of the substances in the mixture were produced in significant quantities. Five of them were listed with yields that ranged from 1.6 to 4 percent. Another eight occurred in amounts that varied from 0.25 to 0.75 percent. I have arbitrarily selected 0.25 percent as the cut-off point for significance, as the importance of the experiment has been identified with the nature of the major products and their limited number. If the cut-off were lowered and a few more products considered, the conclusions that follow would be little affected. What, then, can we learn from this list of thirteen compounds?

A chemist would note immediately that all the above substances belong to a single class of compounds, called carboxylic acids. Amino acids are one subdivision of this class. The result is not entirely surprising, as it is favored by the design of the apparatus. In the discharge chamber, the energy of the spark breaks apart chemical bonds and allows new products to form. No product is safe from further change, however, unless it can escape from the arena. One method of escape lies in the formation of solid, insoluble tars, and most molecules meet this fate. The other route leads out to the boiling-water reservoir. For most organic molecules of small size, however, this haven would be a temporary one. They would reenter the gaseous phase along with water vapor, and be carried back to the spark chamber. Carboxylic acids, however, can find permanent sanctuary in the water flask. Under the conditions of the experiment they are converted to an immobile form on arrival there, and thus are saved.

The family of carboxylic acids is a large one, of course, with an unlimited number of members. Only a select few were made by the Miller-Urey reaction. Which were the favored ones? A glance at the list reveals that those present were all simple ones. The smallest possible carboxylic acid, formic acid, which has only five atoms, was the most prominent product, with a yield of 4 percent. Other substances on the list contained from eight to sixteen atoms. As we noted above, these survivors escaped the spark after limited exposure and had the chance to form only a few chemical bonds. Another factor also reduced the yield of larger molecules. As the number of atoms in a molecule grows, the number of alternative structures that can be built of these atoms increases sharply. No carboxylic acid except formic acid can be made of five atoms, using only carbon, hydrogen, oxygen and/or nitrogen, but three different stable acids have the specific formula $C_3H_7NO_2$ (3 carbon, 7 hydrogen, 1 nitrogen, and 2 oxygen atoms). All three are on our list in the combined total of 2.7%, which, however, must be divided among them. In the case of much larger acids, the number of competing structures would be greater, and the yield of individual ones would diminish accordingly.

Apart from these general considerations, some eccentricities appear in the list. The yields of various carboxylic acids were higher or lower than would be expected if only the above factors were in operation. The second simplest acid, acetic acid, the familiar pungent component of vinegar, has only eight atoms. It was present, but only to 0.5 percent, far behind formic acid. More might have been expected. Such peculiarities reflect the specific chemical processes which take place in the spark, favoring certain pathways and retarding others. Another significant factor also influences the products formed in an experiment of this type, but is less recognized: selection by the experimenter. We can be aware of its influence in this case because Stanley Miller has been quite candid in documenting the course of his work.

His experiment is noted for the production of amino acids, yet in his very first attempt no amino acids at all were detected. He had used the same gas mixture and spark but had placed the various compartments in a different order. Let us continue with his own words: "I filled the apparatus with the postulated primitive atmosphere, water, methane, hydrogen, and ammonia, turned the spark on and let it run overnight. The next morning there was a thin layer of hydrocarbons on the surface of the water, and after several days the hydrocarbon layer was somewhat thicker. So I stopped the spark and looked for amino acids by one-dimensional paper chromatography."

None were found. Miller did not then analyze the nature of the products that had been formed, but rather rearranged his apparatus and tried again. In his next attempt, he obtained a result that satisfied him. This arrangement was then adopted for further work. One modification, tried at a later date, was not helpful. The action of the spark discharge has often been compared to the effect of a thunderstorm. Miller made an effort to improve the analogy: "An attempt was made to simulate a lightning discharge by building up a large quantity of charge on a condenser until the spark jumped the gap between the electrodes. . . . Very few organic compounds were produced and this discharge was not investigated further."

As long as the proper design and components were maintained, however, the same product mixture, including the amino acids, was obtained. Miller took great pains in demonstrating that the products were exactly what he claimed them to be, and that they had been produced by the chemical discharge, not by an accidental introduction of biological material. The overall yields could vary, however. Twenty years after his first studies, Miller wrote: "It was surprising that the yields of amino acids from these first experiments are the highest so far reported in any prebiotic experiment of this type." Thus on his first two tries, he had obtained the worst and the best possible results.

One clear message should emerge from this discussion. A variety of results may be possible from the same general type of experiment. The experimenter, by manipulating apparently unimportant variables, can affect the outcome profoundly. The data that he reports may be valid, but if *only* these results are communicated, a false impression may arise concerning the universality of the process. This situation was noticed by a Creationist writer, Martin Lubenow, who commented: "I am convinced that in every origin of life experiment devised by evolutionists, the intelligence of the experimenter is involved in such a way as to prejudice the experiment."

The Building Blocks

Experiments of the Miller-Urey type have undoubtedly taught us much about the processes of gas-phase organic chemistry. We shall see shortly that they also have significance for cosmochemistry. The question that most concerns us, however, is their relevance to the origin of life. The water, gases, and spark discharge were intended to represent the effect of the sea, atmosphere, and a lightning storm on the early earth. This comparison may not be valid, particularly in the case of the atmosphere. The most important claim usually made for the Miller-Urey experiment, which we encountered in the quote from William Day, is that it produced "life's building blocks." We must pause to remember the identity of these building blocks.

The important construction materials used in a bacterium (or us, if we set aside special equipment such as bones and teeth) are proteins, nucleic acids, polysaccharides, and lipids. Together they make up perhaps 90 percent of the dry weight of a bacterial cell. These large molecules contain from hundreds to billions of atoms. None have been detected, in any amount, in a Miller-Urey experiment. Let us turn, then, to the building blocks of these building blocks. Nucleic acids are made of nucleotides, which themselves are made of a base, sugar, and phosphate. No phosphorus was provided in the Miller-Urey experiments, so no nucleotide could have been formed. A nucleoside (a base-sugar combination, without phosphate) could have been produced, but this did not occur. Nor have any of the dozen sugars, used commonly to construct polysaccharides, or the normal building blocks of lipids been reported in any significant yield in Miller-Urey reactions. Most of these substances contain twenty or more atoms, and would not be expected to occur, for reasons we have discussed.

Finally, we turn to amino acids, the building blocks of proteins. As we have seen, they and other carboxylic acids are the prominent products of Miller-Urey reactions, or at least of those analyzed in any detail. Of the thirteen products formed in highest yield (excluding tar), six were amino acids. Not all amino acids have biological relevance, however. A special set of twenty is employed in biology for the construction of proteins. How well were these represented in spark-discharge experiments?

We start on an encouraging note. Glycine and alanine, members of the set, appear second and fourth on the list, with yields of 2.1 percent and 1.7 percent. However, alanine, and all amino acids other than glycine, appear in two mirror-image forms, with only the L-form relevant to biology. For this reason, only half of the alanine yield has significance. If we were to search the Miller-Urey products for other protein building blocks present in meaningful amounts, we would search in vain. The one formed in the next largest amount occurs, in its L-form, in only 0.026 percent yield (260 parts per million), and the others are even more scarce. They occur among a multitude of organic substances formed in trace amounts, the "Beilstein" mentioned by Urey. The remainder of the substances formed in significant amounts in this experiment cannot be considered building blocks of the large molecules of life. Some appear in lesser roles in one biological system or another. Formic acid, for example, plays some special role in ants. One early method used for its isolation involved the application of dry heat to a flaskful of these unfortunate creatures in the dry state. It would require a stretch of imagination even beyond that often present in this field to see a connection between that event, the prominence of formic acid in a Miller-Urey reaction, and the origin of life.

Let us sum up. The experiment performed by Miller yielded tar as its most abundant product. Of the smaller molecules that were produced, perhaps thirteen may be considered as preferential products. There are about fifty small organic compounds that are called "building blocks," as they are used to construct the four larger types of molecules important to life. Only two of these fifty occurred among the preferential Miller-Urey products. They were glycine and alanine, the two simplest amino acids used in proteins, members of a class that was favored by the design of the experiment. These results have been admirably documented by Miller, and are not in question. It is their interpretation that must concern us.

As we have seen, the reaction product bears no resemblance to the actual content of a bacterium, which is an intricate, organized structure built using large molecules. Even if these large molecules were disassembled into their component parts, the resulting mixture would have only a slight overlap in composition with that of Miller's experiment. The Miller-Urey product has a much greater resemblance to another natural object, however: a type of meteorite.

The Meteorite Connection

Not all of the debris present in the solar system at the time of its formation was captured by the sun, the planets, and their satellites. A number of smaller fragments survived in independent orbits. The rocky ones are called meteorites, while others, made mostly of ice, we know as comets. On occasion, a meteorite may enter our atmosphere and survive to strike the surface of the earth. The recovered fragments have been subjected to intense study, as they are samples of the original material present in the solar nebula 4.5 billion years ago and perhaps can tell us something of the origin of our solar system. Meteorites may even contain particles of interstellar material that predate our solar system. These topics are fascinating but do not concern us here. Our interest lies in a subclass of meteorites called carbonaceous chondrites, which contain a small percentage of carbon.

Most of this carbon is bound into a tarry, insoluble material. The remainder consists of a very complex mixture of smaller molecules which has been called "a random suite" or "a chemical stockroom" by members of the scientific teams who performed the analyses. The term "Beilstein" could be used as well, as each component is present in a very limited amount. Carboxylic acids of various kinds are present, including many amino acids. When a comparison is made of the identity and *relative* amounts of amino acids present in these meteorites and those in the Miller-Urey experiments, striking similarities are observed. We will cite two of the scientists, J. G. Lawless and E. Peterson, directly: "Comparison of the linear neutral amino acids present in the Murchison meteorite, in laboratory chemical evolution experiments, and in a terrestrial organism shows a marked similarity between the meteorite and laboratory experiments and a significant difference between the meteorite and *E. coli.*"

The Murchison meteorite is a much-studied object that fell in Australia in 1969, and *E. coli* is the short form of the name of an even more studied bacterial strain, *Escherichia coli,* which inhabits our guts. The Miller-Urey experiment, then, may have modeled some of the processes that took place among the reduced gases in the original solar nebula to form compounds now preserved within meteorites. I have used the word "some" just now and emphasized "relative" in the above paragraph because amino acids and other carboxylic acids occur in much lower absolute amounts in meteorites than in Miller-Urey experiments. As suggested earlier, the design of the spark apparatus may have favored these compounds and enhanced their yields relative to the amounts that would be expected in an appropriate natural situation. If we set aside this enhancement, the lasting contribution of these experiments may be as a model for certain chemical processes in outer space.

The Realm of the Predestinists

We are left with an unresolved, enigmatic question that involves psychology and history more than chemistry: Why has the Miller-Urey experiment had such a strong impact on the origin-of-life field? To answer this, we must look at a number of different belief systems.

In the early nineteenth century, it was felt that the crucial distinction between living and nonliving systems lay in the nature of the very chemicals used to construct them. Organic compounds contained the life force, while inorganic substances did not. The name "organic chemistry" was meant to describe the field now called biochemistry. In 1828 a German chemist, Friedrich Wohler, prepared urea, a component of urine, from another substance that was considered inorganic. Wohler wrote to a colleague: "I must tell you that I can prepare urea without requiring a kidney or an animal, either man or dog." Since that time it has been recognized that the preparation of organic compounds is a feat of no profound difficulty, nor one of any great significance to life. The discovery of mixtures of organics in meteorites and, as we shall see, in interstellar space demonstrates the ease and universality of this process. The difficult step in the origin of life lies further down the line, not here. Yet some of the confusion in the media and in science may concern just this point. Chemist William Day states in his description of the Miller-Urey results: "No longer was there the dilemma of how organisms could have produced organic compounds before they themselves existed—the building blocks had already been there on the primordial earth." That dilemma had been laid to rest a century earlier.

Confusion also exists concerning the actual products of the experiments. Miller has certainly been forthright and accurate in his publications and summaries. Yet we find the following statement in A. L. Lehninger's widely used textbook, *Biochemistry*: "Many different forms of energy or radiation lead to organic compounds from such simple gas mixtures, including representatives of all the important types of molecules found in cells as well as many not found in cells." That statement, as written, is simply incorrect. For some molecules it is true, if one ignores considerations of yield and attributes significance to the mere presence of a substance, in whatever amount. Recently, for example, Cyril Ponnamperuma detected the five bases used in DNA and RNA (which contain from twelve to sixteen atoms each) in both a Miller-Urey type of mixture and a meteorite. The compounds occurred to the extent of perhaps 2 parts per million, yet Ponnamperuma in a news conference called it "almost an awesome result." The awe must lie in the eye of the beholder. Nothing within the result compels it.

Other biochemical substances, nucleosides for example, have never been reported in any amount in such sources, yet a mythology has emerged that maintains the opposite, and extends this conclusion to even more complicated molecules. I have seen several statements in scientific sources which claim that proteins and nucleic acids themselves have been prepared by subjecting a reducing atmosphere to various energy sources.

These errors reflect the operation of an entire belief system, one that I call predestinist. A predestinist believes that the laws of the universe contain a built-in bias that favors the production of the chemicals vital to biochemistry and ultimately to human life itself. No difficult or extended process was involved in the origin of life, according to this system. If we set up the right experiment, everything would fall quickly into place. To a predestinist, the Miller-Urey experiment provided the expected validation of his beliefs. If glycine and alanine were present, surely the remaining amino acids would also appear in large quantities, and nucleotides too, as soon as the appropriate experimental modifications were made. The principle had been proved; the remainder was just a matter of mopping up.

The facts do not support the belief, nor can we extrapolate it from what we know. The nucleotides, for example, when built into DNA, perform the task of information storage and transfer quite well. Presumably, a long period of evolutionary trial and error was needed to develop this mechanism. Why would we expect the necessary components to be made preferentially on the early earth, before the start of life? If this was the case, then obviously someone had arranged things that way. This bias might be due to a mystical spirit of cosmic evolution or to an actual deity. Someone or something out there cares about us. Such thoughts may be comforting, but they run far ahead of any experimental validation. They belong to religion or mythology, not to science.

The Oparin-Haldane Hypothesis

Thus far, we have discussed the first Miller-Urey experiment as if it had occurred in isolation. In fact its inspiration and its impact were connected to the historical circumstances that preceded it. The experiment was accepted as validation not only for the beliefs we have described, but for a larger theory that had gradually come into scientific favor. As we mentioned earlier, this theory was put forth independently in the 1920s by Alexander Oparin in the Soviet Union and J. B. S. Haldane in England.

Their hypothesis filled a near vacuum in thought on the origin of life that had existed since the demise of spontaneous generation. Pasteur had demonstrated that living beings arose only from earlier living beings. How, then, did the first life arise? In the absence of a viable scientific answer, those needing a solution could only turn to religion. To some scientists, particularly those defending evolution from attack by fundamentalists, this situation was unacceptable. The most obvious remedy was the revival of spontaneous generation in some form, with the added provision that it required conditions that were present long ago on earth, but not now. In addition, the thought arose that the formation of an entire bacterium might not be necessary. To start life, it might suffice if some smaller part of a cell, a protein or even a bit of gel-like protoplasm (cellular fluid), came into being.

Haldane published his ideas once, then turned his attention to other areas of science. Oparin, however, continued to expand his theory. It came to greater scientific attention when his book was translated into English in 1938, and gained prominence and credibility when Harold Urey endorsed and extended it in the early 1950s. Urey had won the Nobel Prize in chemistry in 1934 for his discovery of a new stable isotope of hydrogen called deuterium. During World War II he played an important role in the Manhattan Project, which developed the military uses of atomic energy. Subsequently he took a strong interest in the chemistry of the solar system. In his influential book *The Planets* (1952), Urey supported the various components of the Oparin-Haldane hypothesis.

In its mature form, this theory can be summarized as follows: (1) The earth, at the time when life began, had a reduced oxygen-free atmosphere, with methane, ammonia, hydrogen, and water. (2) This atmosphere was exposed to various energy sources, such as lightning, solar radiation, and volcanic heat, which led to the formation of organic compounds. (3) These compounds, in Haldane's words, "must have accumulated until the primitive oceans reached the consistency of hot dilute soup." (This last word has stuck in the public mind, and the oceanful of organic substances is now generally called the prebiotic or primordial soup. A recent display in the NASA Aerospace Museum in Washington showed a film of television chef Julia Child preparing such a soup. For reasons to be explained later, I would not recommend it for human consumption. Some bacteria, however, would thrive on it. For this reason, and because of the oxygen in the air, such a soup could not persist today.) (4) By further transformations, life developed in this soup. According to Urey, the soup components "would remain for long periods of time in the primitive oceans ... this would provide a very favorable situation for the origin of life."

The theory did not specify the details of this last step. As we shall see, considerable disagreement exists on this topic; Haldane and Oparin themselves had very different ideas concerning it. We shall need much of the remainder of this book to sort out the various possibilities. For now, we shall focus on the first three parts of the theory, since they constitute the reigning paradigm on the origin of life.

In considering them we shall need the attitude of the Skeptic, to sort out logic from illogic and science from mythology. To start, we must note that the Miller-Urey experiment was inspired by the entire theory, but tested only the second point. Yet the assumption is often made that the entire theory was confirmed by it. For example, a current geology text by R. A. Goldsley states: "These experiments have produced many of the chemicals fundamental to life. It seems likely, in view of these results, that Haldane's description of the earth's primeval oceans as a 'hot dilute soup' of organic molecules was correct." Yet this description would be accurate only if points 1 and 3 received independent confirmation. In fact, the evidence and the opinions of those scientists most concerned with the subject have moved in the opposite direction.

A Change in the Air

The presence of a strongly reducing atmosphere is a central assumption of the Oparin-Haldane hypothesis, and underlies the design of Stanley Miller's experiment. Of course, we have no way to sample the air of 4 billion years ago, and inferences concerning its composition must be indirect. Urey based his argument on the cosmic abundance of hydrogen and the probable composition of the solar nebula. As we discussed in a previous chapter, the current geological consensus supports the idea that the atmosphere came from the interior of the earth rather than the nebula. Thoughts concerning its composition vary, but the most frequently heard guess supports the presence of nitrogen, carbon dioxide, water vapor, and a bit of hydrogen, but no methane, ammonia, or oxygen. This atmosphere is neutral for the most part, with a slight reducing power. Geologists now realize that a methane and ammonia atmosphere would have been destroyed within a few thousand years by chemical reactions caused by sunlight.

Stanley Miller and others have attempted to prepare amino acids under the new conditions. The ratio of hydrogen (H_2) to carbon dioxide (CO_2) is a crucial variable. When this falls below 1, as the above example specifies, only glycine is produced, in trace amounts, but no other amino acid. Miller has been quite frank in his statements: "There are difficulties in maintaining H_2/CO_2 ratios greater than 1.0 [for the early earth] because of the escape of H_2 from the atmosphere. Adequate sources of H_2 to maintain this ratio are possible but difficult to justify." Elsewhere he notes: "If it is assumed that amino acids more complex than glycine were required for the origin of life, then these results indicate a need for CH_4 [methane] in the atmosphere."

It is the Oparin-Haldane hypothesis, actually, that requires methane in the atmosphere. If this gas or other reducing substances were absent, it would mean that some other course of events, not described by the theory, led to the origin of life. This distinction has been missed, however, by some supporters of the hypothesis. The astronomer Manfred Schidlowsky stated at a 1977 meeting, for example: "The very fact that life sprang up on Earth constitutes conclusive proof of a primary reducing environment since the latter is a necessary prerequisite for chemical evolution and spontaneous origin of life." And a 1983 biochemistry text edited by Geoffrey Zubay contains the following statement: "The primitive atmosphere must have contained reducing equivalents in some form to yield amino acids, since no biomolecules or their precursors are formed when a mixture of carbon dioxide, water, and nitrogen is sparked."

We have reached a situation where a theory has been accepted as fact by some, and possible contrary evidence is shunted aside. This condition, of course, again describes mythology rather than science.

The Myth of the Prebiotic Soup

The prebiotic soup has fared little better than the reducing atmosphere. The above section heading is not my own, but was taken from an important article by a Swedish geologist, Lars

Gunnar Sillen. He starts with the assumption of a methane-rich reducing atmosphere, but questions the survival of a soup under these conditions. If left to itself, he reasons, it would gradually move to the position of greatest stability, equilibrium. When this position was attained, we would be back at the starting point, with almost all carbon in the form of methane and concentrations of amino acids at negligible levels. A system may be kept away from equilibrium, of course, by a steady input of energy. All life exists in this situation today. Enormous amounts of energy would be required, however, to maintain an entire ocean in this condition. Furthermore, mixtures of organic chemicals are far less adept than living systems at handling a heavy energy flow. As we saw in the Miller-Urey experiment, they continue to form chemical bonds until a heavy insoluble material, a tar, is produced, unless they are protected in some type of sanctuary.

Some evidence from our present world supports this concept. A certain amount of biological material that is released into the oceans is altered by random chemical events so that it is no longer palatable to living organisms. This material may then serve as a model for all of the organic substances present in the ocean before life began. Chemist Arie Nissenbaum has studied its fate, and noted that it does not accumulate in the oceans. Concentrations remain quite low, and the average age of the material is no more than 3,500 years. It is depleted by a number of geological processes. Heavier molecules settle out and form deposits. Other substances are absorbed by minerals, which compact into sediments. The sediments deposited throughout the entire record of geological history contain organic components of this type. A prebiotic soup, if ever formed, would presumably suffer the same fate, before it had the chance to meet its alternative destiny, the return to equilibrium.

The Retreat from the Hypothesis

Awareness of these developments has spread in recent years in the scientific community concerned with the origin of life and caused erosion of the paradigm, the Oparin-Haldane hypothesis. As one would expect in this situation, efforts have been made to save whatever features could be rescued. There have been speculations that sufficient organic material to stock the prebiotic oceans could be brought in by meteorites, by comets, or even by an encounter with a cosmic dust cloud. Such suggestions would abandon the reducing atmosphere, but save the soup. To provide enough material to support schemes of this type, special assumptions must be made about the rate of infall of extraterrestrial bodies and the survival of organic compounds during the process of entry and impact. No independent evidence supports these assumptions. Such speculations cannot be discarded, but must be held in abeyance until some confirmation arrives. One related scheme that involves comets is so spectacular, however, that we later will give it a chapter of its own.

Another dramatic and picturesque alternative has been provided by the biologist Carl Woese, of the University of Illinois. Professor Woese has been outspoken in his criticism of the current dogma, stating: "The Oparin thesis has long ceased to be a productive paradigm: It no longer generates novel approaches to the problem; more often than not it requires modification to account for new facts; and its overall effect now is to stultify and generate disinterest in the problem of life's origin. These symptoms suggest a paradigm whose course is run, one that is no longer a valid model of the true state of affairs."

Woese's own proposal is certainly novel. He suggests that life began in the earliest days of earth, before the planet was fully formed. The mantle, core, and crust were not entirely sorted out at that time. Large amounts of metallic iron were present at the surface. It entered into chemical reactions that produced an atmosphere with carbon dioxide and hydrogen. Enough carbon dioxide was present to cause a "runaway greenhouse effect," torrid conditions similar to those on Venus today. The surface was hot, and perhaps molten in places. Meteor infall was heavy. Strong winds produced violent dust storms, carrying particles high into the atmosphere. Water vapors condensed on this dust, producing vast clouds of tiny water droplets. These clouds, the sole inhabitable oasis on a turbulent planet, served as the cradle of life. Each droplet acted as a primitive cell, a small laboratory for experiments in chemical evolution.

According to this theory the atmosphere and dust provided raw materials, and sunlight furnished energy. The first organisms to evolve were the methanogens, which reduced the carbon dioxide in the atmosphere by combining it with hydrogen. As carbon dioxide levels fell, the runaway greenhouse conditions abated and the earth cooled down. Oceans could then form, and our planet approached its present state.

Less sweeping solutions have also been proposed to get around the difficulties of the current paradigm. If a reducing environment is needed for the origin of life, one need not divert an entire planet to that purpose. It would suffice to have some local niche where reducing conditions prevailed. Charles Darwin himself suggested a small pond for the origin of life; others have followed his example. Tidal pools have been another popular alternative. The most fashionable place within the last several years has been a very different location, hot vents at the bottom of the sea.

Such vents occur at places where the earth's crust is thin and molten rock penetrates close to the surface. A group of them occur near the Galápagos Islands, the place where Charles Darwin obtained a number of insights into the origin of species. This site has been explored intensively, in expeditions that used a submersible vessel, the *Alvin*.

The vents emit reduced chemicals, including hydrogen sulfide, methane, and ammonia, in addition to hot water. Bacteria live on the chemical energy of hydrogen sulfide, while more advanced organisms such as worms, mussels, and clams ultimately depend on the bacteria for food. Thus an entire ecology exists at the bottom of the ocean, independent of sunlight.

Water boils at elevated temperatures at the high pressures that exist under 2,500 meters (8,000 feet) of ocean. Remarkably, some bacterial communities appeared to thrive at temperatures of 360°C (680°F) under these conditions. Taken to the laboratory, specimens grew under pressure at 250°C (482°F). Previously, no known microorganism had been known to survive for long when held above 105°C (221°F). A remarkable report of this type of course requires a skeptical response from other scientists. Replication and confirmation will be needed for ultimate acceptance. Claims have in fact been made that the results are artifacts, and the resolution is uncertain at this time.

These unusual circumstances, together with the reducing conditions that prevail in the waters of the vents, have stimulated speculation that life arose in such vents early in the history of the earth. Chemical processes would proceed at advanced rates in this environment, independent of events in the atmosphere. The limited volume and relatively brief lifetime of individual vents, on the other hand, provide adverse factors. The vents are a possible location for the origin of life, but not the only one, nor need they be the one that is most favorable.

Many other local solutions to the problem of a reducing environment may turn up. We cannot even be sure, at this stage, that a reducing environment was necessary. The clay theory, for example, holds a different point of view. Moreover, the specification of the correct location is not the most critical problem faced by origin-of-life theory. The very best Miller-Urey chemistry, as we have seen, does not take us very far along the path to a living organism. A mixture of simple chemicals, even one enriched in a few amino acids, no more resembles a bacterium than a small pile of real and nonsense words, each written on an individual scrap of paper, resembles the complete works of Shakespeare. The subsequent events that occur within the initial chemical mixture are the ones that concern us. These events were not in the scenarios that we discussed in this chapter. Popular accounts of the origin of life usually give little attention to this aspect of the problem. It is assumed that if enough time is provided, the random shuffling of molecules within the prebiotic soup will sooner or later produce a living system. We will give this assumption our skeptical attention in the next chapter.

five

The Odds

A number of colorful and exotic places have been suggested for the origin of life: the clouds, the bottom of the sea, tidal pools, comet interiors, and alien planets circling other star systems. These suggestions have been so spectacular that they have caused the problem of the site of the origin to overshadow a more fundamental question: What process was involved when life originated?

The advocates of each location have usually argued that their site is the most appropriate one for Miller-Urey type chemistry. The proper reducing environment would be present there, and the reactions would work as well as they do in the laboratory. But even if they did so, little would have been accomplished. An immense gap separates a chemical mixture that contains a few amino acids and the highly organized complexity of the simplest cell alive today.

The smallest free-living organisms are probably the mycoplasmas, tiny bacteria that are only a fraction of the length of the more typical one that we saw on our COSMEL trip. At the -6 level of COSMEL, where bacteria such as *E. coli* are about our own size, a small mycoplasma would approximate the size of a basketball. Yet these tiniest creatures still possess cell membranes, ribosomes, DNA, hosts of enzymes, and the other complexities associated with all life on this planet. As we shall see, viruses are generally smaller than mycoplasmas, but they are not separate living beings. They function as parts of organisms, not as complete ones.

If life originated from a simple chemical mixture, then we want to know the steps that led from this mixture up the ladder of organization to the first cell. This same question would remain whether the mixture were formed in some environment on earth or anywhere else in the universe. We have seen that replication and natural selection provide a reasonable mechanism for the further evolution of the common ancestor. This creature, however, may have been close to a bacterium in its complexity. Unfortunately, we are uncertain about the processes that produced it.

One common presumption throughout history has been that the first organism was formed by chance. An appropriate mixture rearranged itself at random until a living cell emerged. Such thoughts were widely held up to the time of Louis Pasteur, when the complexity of even the smallest cells was not yet recognized. His experiments provided convincing proof against the spontaneous generation of bacteria. Yet the concept died slowly. Decades later, at the start of the twentieth century, Henry Bastian was still at work in his laboratory in hopes that the application of the right amount of heat would suffice to kill all living things in his broths, yet leave these infusions with the capacity to generate new life. In a later chapter we shall meet Olga Lepeshinskaia, who was awarded a Stalin Prize in 1950 for her work describing the spontaneous generation of cells. However, these responses are anomalies, and virtually all scientists today believe that living cells cannot commonly be generated from their chemical ingredients by random processes.

Creationists, and some other religious groups, occasionally cite this situation as evidence for the existence of God. One favorite analogy involves the discovery of a watch on a walk in the wilderness. Imagine that we found one, in working condition, and on looking inside, saw a bewildering array of gears and springs which served to keep the various hands moving smoothly along their course. We would not presume that this mechanism had come together from its parts by chance. It would function only if its components had been put together in exactly the right way, by a watchmaker. Similarly, the existence of bacteria and other living beings, all of which are much more complex than a watch, implies the existence of a creator, as only a higher being could design creatures so fit for their function.

We will not take this escape route in our book, for we are committed to seeking an answer within the realm of science. If a watch is complex, then the watchmaker must be even more complicated. A being with the capacity to create a watchmaker would be the most complex of the lot. By following this line of reasoning, we have made our problem more difficult rather than simpler, and we can resolve it only by introducing supernatural forces. We must look for another solution if we wish to remain within science.

The watch analogy serves to introduce us to the nature of our problem, but understates it. It would not suffice to assemble a watch by chance, by shaking its parts together in a box, to mimic the spontaneous generation of life, for the parts themselves are manufactured items. Spontaneous generation calls for the assembly of a functioning cell from the raw materials in the environment. As an approximation to this process, we must imagine that we put a suitable quantity of crude mineral ores in a box and shook them together. The ores would include iron and other metals, silicates (for the glass), and limestone (to provide carbon for the diamond bearings). If these ores, shaken together, rearranged their atoms to form a watch, then we would have carried out a more exact imitation of spontaneous generation.

Even this exercise would not capture the actual situation. In the above example, we intervened by selecting the ores, placing them together, and shaking them to help them interact. To eliminate this intervention, we should search for a place in the wilderness where the appropriate ores lay in suitable proximity to one another. If natural processes such as lava flows, falling rocks, flowing water, and earthquakes then served to gather and refine the ores, shape them into parts, bring the parts together, and assemble them into a functional watch, *then* we would have observed a fitting analogy to the spontaneous generation of a bacterium.

But bacteria differ in a vital way from watches. One procedure exists which can convert a Miller-Urey type of mixture into bacteria. We need only add a single bacterium of an appropriate type to the chemicals, and wait. After a few days, a large number of new bacteria will have been created, using the materials in the mixture. The transformation has been demonstrated recently using a material called tholin as a food supply. This complex organic solid, a product of spark discharges in certain reducing atmospheres, is related to the tars formed in the Miller-Urey experiment.

When appropriate chemical food is provided, billions of bacteria can be produced from a few, within days. Yet the entire process would not take place without the initial seed. The process of multiplication is so dramatic, however, that some scientists who reject spontaneous generation as a common occurrence are tempted to invoke it just once in the history of the earth, to get life under way. Given that one event, all else could follow, and the problem of our origin would be solved.

Professor George Wald is perhaps the most eloquent representative of this point of view. He is a Harvard biochemist who won the Nobel Prize in 1967 for his studies in the chemistry of vision. He has also been outspoken on a number of topics outside his area of achievement, including the origin of life. His comments on spontaneous generation published in an article in *Scientific American* in 1954 have been cited widely in texts and anthologies. I will continue this tradition, and quote him directly here: "One has only to contemplate the magnitude of this task to concede that the spontaneous generation of a living organism is impossible. Yet we are here—as a result, I believe, of spontaneous generation."

This contradiction is resolved by revising our concept of the impossible. Professor Wald points out that we tend to use this word to apply to events that are very unlikely in our daily experience. If repeated trials of an event can be made over a very long time period, however, one that is much longer than human history, the odds will improve considerably.

We can show how repeated trials make an improbable event into a likely one: Imagine a box containing ten coins. If we shook the box and looked to see how the coins had landed, the chances of getting all heads would be less than 1 in 1,000. It is quite unlikely that this would occur in a single trial. But suppose that we could shake the box 1,000 times. The chance that we would get ten heads at least once is now 63 percent. It has become probable.

Wald points out that we are not familiar with the idea of very large numbers of trials. Given sufficient trials, however, very improbable events become likely. I can present another illustration of this. A national lottery may have odds of 10 million to 1. If we were to win such an event we would consider ourselves very fortunate. Yet if we could buy only one ticket each day, but were able to continue this habit for 30,000 years, a win would become probable. (Unfortunately, our winnings would probably not pay for our cumulative expenditure on lottery tickets!)

In the case of the origin of life, a single win would be sufficient. The time span involved could be a billion years, and the entire surface of the planet would be available for trials, so that many could be run simultaneously. Let us quote geologist R. F. Flint in his text *The Earth and Its History:* "How many times 10,000 trials of such random events could have occurred within a period of 3.3 billion years? One's imagination boggles at the thought of trying to calculate so great a number. No one familiar with statistics rejects the idea of chance chemical combinations because there wasn't enough time. There was a huge abundance of it."

For another statement of this view we can return to George Wald's *Scientific American* article: "Time is in fact the hero of the plot. The time with which we have to deal is of the order of two billion years. What we regard as impossible on the basis of human experience is meaningless here. Given so much time, the 'impossible' becomes possible, the possible probable, and the probable virtually certain. One has only to wait: time itself performs the miracles."

Thus the vast improbability of spontaneous generation is

countered with the immensity of the surface of the earth and of available time. It would be hard to find a more eloquent statement of the above point of view than Professor Wald's, but is it correct? We want to weigh the argument and measure the quantities involved, rather than be overwhelmed by them.

To start, we must refuse to let our minds be boggled by huge quantities. Mathematician Douglas Hofstadter has written about the inability of many people to deal with very large numbers, such as those used in defense spending or astronomical periods of time. He asks whether "we actually suffer from number numbness. Are we growing ever number to ever bigger numbers?" He later calls this condition "innumeracy," the mathematical equivalent of illiteracy.

We will not be able to evaluate Professor Wald's argument if we suffer from this condition, for we must compare some very large numbers. We will want to know the odds against success on the one hand, and the total number of trials that we can make, on the other. If the number of trials is much greater than the odds, our prospects are good, but if the odds are greater, our chances are poor. To learn which situation applies in the case of spontaneous generation, we must estimate both quantities and compare them.

In comparing very large numbers, our everyday English serves us poorly. A recent *National Geographic* article described the energy released by a quasar in the following way: "Consider a big nuclear power plant producing 1000 megawatts of electricity. Multiply those 1000 megawatts by a billion trillion. Then multiply again by 10 billion." Now how much energy is that? Surely, that amount would keep New York City lit up for a while. But is it larger, say, than a trillion trillion trillion megawatts? We will need a better system for the work we have ahead of us.

Scientists avoid cluttering up a page with zeros by using a system called exponential notation. The number 10 is written, with some numeral set to the right and above it, for example 10^3. This number can be converted to an ordinary one simply by writing down "1," followed by the number of zeros in the upper numeral. In the case of 10^3, we would write "1,000." It is not difficult to write 1,000 in our usual way, but when we get to very large numbers the exponential system is very valuable. It is far easier to write 10^{18} than 1,000,000,000,000,000,000. It is not instantly obvious that the last number is larger than 100,000,000,-000,000,000, but we can tell at once that 10^{18} is larger than 10^{17}.

Using this exponential notation, much larger periods of time than that involved in the evolution of life on earth become manageable. I was not troubled by an article I read recently that projected a possible future history of our universe according to "open" models: All stars will have run out of fuel and cease to shine after 10^{14} years or so. They will lose all their planets through close encounters with other stars in 10^{17} years. All protons will have decayed by the year 10^{32} and matter of the types familiar to us will have ceased to exist. Finally, by the year 10^{100} black holes will have lost their mass by evaporation.

A new problem can arise, however, in dealing with numbers of this magnitude. In considering this possible future history, I found myself assuming that the time needed for the stars to lose their planets would be only 20 percent greater than the time until they ran out of fuel. I mentally compared the exponential 14 and 17 as if they were numerals, rather than the number of zeros following a 1. In fact 10^{17} is 1,000 times larger than 10^{14}. If we considered the history of the universe at the time when the planets had been lost, then the stars would have been shining for only the first 0.1 percent of that history. They would have been extinguished for all the remainder of that time.

I will introduce a new device, the Tower of Numbers, to help us in dealing with these large figures. Like COSMEL, this tower is logarithmic—at each level, things are ten times larger than they were on the level just below. Unlike COSMEL, however, the Tower of Numbers has no elevator, only a staircase. The word "tower" was chosen to give it a sound of antiquity, and also suggests the biblical Tower of Babel, which was intended to reach unto heaven. The Tower of Numbers, unlike COSMEL, extends upward indefinitely.

This device can be used to keep count of any type of object, but for our first example, let us pick a very familiar one, money, in the form of pennies. If we enter, in our imagination, at ground level we would find a room in which the floor is covered with pennies. The supply is inexhaustible, for as we pick some up, additional ones are delivered through an opening in one wall. The only other things to be seen in the room are a staircase that leads up to the next floor and a display counter with an attendant. The counter contains items that can be bought for one to nine pennies. For example, the attendant will sell us two aspirins or four raisins or five toothpicks for one penny, and we can buy a mint candy for six pennies.

If we want more expensive items we must go up to a higher level. We climb up one flight and reach the first floor (the European system of naming floors is in use). Again there is an attendant and a display counter, with items now priced from 10 to 99 cents. This attendant prefers to be paid in dimes. Unfortunately, there are no coins to be found on the floor here. He will, however, gladly give us a dime in exchange for ten pennies that we bring up from below. The essence of this story will not be affected by the number of coins that we can manage to lift at once, but to simplify things, let us suppose that there is a rule that only ten coins may be carried up the staircase on any trip. To buy a newspaper (cost: three dimes), we would have to climb from the ground floor to the first floor three times, carrying ten pennies each time.

The point is that as we go to each new floor, it becomes progressively harder to buy anything. Dollars are accepted at the counter on floor two. To get a dollar, however, we would have to climb ten times from the ground floor to the first floor, to obtain ten dimes. These coins would then be carried to the second floor, to exchange them for a dollar. If we wished to purchase a bottle of wine for three dollars, we would need to repeat this process two more times.

If our time and energy were unlimited, we could eventually carry cash up to the higher floors and purchase a bicycle on the fifth floor, a car on the seventh floor, or a house on the eighth floor. Going higher yet, we would find the annual budget of the United States government on the thirteenth floor and the gross national product of the country on the fourteenth. It is likely that we could purchase all of the property on earth for the number of pennies needed to take us up to the seventeenth floor.

The logarithmic construction of our tower works ever more ferociously against us as we continue to climb. If we wished to buy a house (eighth floor), and had saved enough money to reach the seventh, then we would not be almost there, in spite of the fact that we had traveled seven-eighths of the way up. We would only have $10,000, and need $90,000 more. If we had worked our way up from the penny level, we would now have to return there and repeat our entire effort nine more times.

Our tower can represent commodities other than money. Atoms, for example, are more relevant to the theme of this book. Assume that the ground floor was filled with atoms, of all types, in unlimited supply. If we selected two hydrogen atoms and one oxygen atom, we could exchange them at the counter for a water molecule. With nine atoms (two carbon, six hydrogen, one oxygen), we could purchase a molecule of ethyl alcohol. Only very simple organic molecules can be obtained on the ground floor.

To get the components of life, we would have to climb higher up. On the first floor (10–99 atoms) we would find amino acids, nucleotides, and simple sugars. Most lipids would be available on the second floor, while enzymes and RNA molecules would be found on the third and fourth floors. If we wanted to obtain the DNA double helix which makes up the chromosome of a bacterium, we would have to climb to the eighth floor, while a ribosome would be found slightly lower, on the seventh. The construction of an entire bacterium would require enough atoms to take us to the eleventh floor, while a trip up to the twenty-seventh floor would be needed to obtain a human being. Continuing up, we would encounter the earth on floor 51 and the sun on floor 57. The universe would be in the display case on floor 78 (it has perhaps 10^{78} atoms).

We are now ready to handle the chances for the spontaneous generation of a bacterium. Using the Tower of Numbers to estimate trials rather than pennies or atoms, we can put the "mind-boggling" number on its proper level. For our purposes, we will want to overestimate and select the largest number of random trials that might have been attempted on the early earth, as the actual number would be very difficult to determine.

We need to know two items, the length of time needed for a single trial and the number of trials that can take place simultaneously. Under the most favorable conditions, an *E. coli* colony can double in about twenty minutes. In other words, it takes twenty minutes for a bacterium to assemble a replica of itself from simple chemicals. It is unlikely that a bacterium would come together more quickly by random processes. Let us presume, however, that a simpler bacterium than *E. coli* is involved, and estimate one minute as the time for a trial. If we accept the evidence of the fossils and the usual age cited for the solar system, then a maximum of 1 billion years, or 5×10^{14} minutes, was available for the origin of life on earth.

What about available space? As a maximum estimate, we can assume that the entire earth was covered by an ocean 10 kilometers deep, which was available for experiments. Further, we will allow that space to be divided into small compartments (1 micrometer on each side) of bacterial size. We would then have 5 times 10^{36} separate reaction flasks. If a separate try was made in each flask every minute for 1 billion years, we would have 2.5 times 10^{51} tries available. We would be on the fifty-first floor of the tower.

That is a very large number, and we are probably several floors too high in our estimate, but we will use it, to continue the argument. Is it large enough to justify any event whatever? The Skeptic would disagree. Some unlikely occurrences will become probable, given this number of trials, but others will not. If we recall the case when we flipped ten coins at once, we were unlikely to get ten heads on the first attempt, as the odds were 1,000 to 1 against us. The result became probable, however, when we had 1,000 trials at our disposal. As a rough rule, we will consider that an event becomes probable when the number of trials available is of the same order of magnitude (falls on the same floor of the tower) as the adverse odds on a single trial. In the case of spontaneous generation of a bacterium, if these odds are represented by a number that falls much above the fifty-first floor, then it is very improbable, even though we had a large number of trials at our disposal.

We cannot compute these odds precisely, but approximations

will serve our purposes quite well. Many scientists have attempted such calculations; we need cite only two of them to make the point. The first was provided by Sir Fred Hoyle, whose ideas we shall discuss in detail later in the book. He and his colleague, N. C. Wickramasinghe, first endorsed spontaneous generation, then abruptly reversed their position. Why did they do this? Quite obviously, they calculated the odds.

Rather than estimate the chances for an entire bacterium, they considered only the set of functioning enzymes present in one. Their starting point was not a complex mixture, but rather the set of twenty L-form amino acids that are used to construct biological enzymes. If amino acids were selected at random from this set one at a time and arranged in order, what would be the chances that this process would produce an actual bacterial product? For a typical enzyme of 200 amino acids, the odds would be obtained by multiplying the probability for each amino acid, 1 in 20, together 200 times. The result, 1 in 10^{120}, places us on floor 120 of the Tower of Numbers, immensely higher than the level where we find the number of trials.

Things need not be that bad, however. What matters is the function of the enzyme, rather than the exact order of amino acids within it. A large number of amino acid sequences might provide enzymes with the proper function. With this in mind, Hoyle and Wickramasinghe estimated that the chances of obtaining an enzyme of the appropriate type at random were "only" 1 in 10^{20}. To duplicate a bacterium, however, one would have to assemble 2,000 different functioning enzymes. The odds against this event would be 1 in 10^{20} multiplied together 2,000 times, or 1 in $10^{40,000}$. This particular item would then be available on floor 40,000 of the Tower of Numbers. If we consider that the number of trials brought us only to the fifty-first floor, we can understand why Hoyle changed his mind. His estimate of the likelihood of the event was that it was comparable to the chance that "a tornado sweeping through a junk-yard might assemble a Boeing 747 from the materials therein."

In fact, things are much worse. A tidy set of twenty amino acids, all in the L-form, was not likely to be available on the early earth. This situation has not even been approached by the very

best Miller-Urey experiments. Nor does a set of enzymes constitute a living bacterium. A more realistic estimate has been made
by Harold Morowitz, a Yale University physicist. He has calculated the odds for the following case:

Suppose we were to heat up a large batch of bacteria in a
sealed container to several thousand degrees, so that every chemical bond within them was broken (the same transaction could be
performed with our imaginary Atom Grinder). We then cooled
this mixture slowly, in order to allow the atoms to form new
bonds, until everything came to equilibrium. In this state, the
most stable chemicals (those with the least energy) would dominate the mixture, while those with higher energy would be present to a lesser extent, in accordance with the laws of statistics.
Morowitz asks, what fraction of the final product will consist of
living bacteria? Or in other words, if a single bacterium was used
to start the experiment (ensuring that the appropriate atoms, in
proper amounts, were present), what would be the chances that a
living bacterium would result at the end?

The answer computed by Morowitz reduces the odds of Hoyle
to utter insignificance: 1 chance in $10^{100,000,000,000}$. We are on the
100 billionth floor of our tower! This number is so large that to
write it in conventional form we would require several hundred
thousand blank books. We would enter "1" on the first page of
the first book, and then fill it, and the remainder of the books,
with zeros. If, by some unimaginable method, we were to obtain
enough trials to ascend in our tower to floor 99,999,960,000, then
we would face "only" the odds cited by Hoyle.

The Skeptic will want to rewrite Professor Wald's conclusion:
Improbability is in fact the villain of the plot. The improbability
involved in generating even one bacterium is so large that it reduces all considerations of time and space to nothingness. Given
such odds, the time until the black holes evaporate and the space
to the ends of the universe would make no difference at all. If we
were to wait, we would truly be waiting for a miracle.

One escape hatch yet exists for spontaneous generation. Why
need the event have been probable? We can just stare at the
odds, shrug, and note with thanks how lucky we were.

After all, improbable events occur all the time. For example, the odds in the lottery we cited were 10 million to 1. As we have seen, we would have to buy a ticket every day for about 30,000 years to make a win a probable event. Yet every so often we note in the newspapers that there is a winner. That person was not 30,000 years old, and usually has bought just one or a few tickets. He was merely lucky.

If I wish, I can create a rare event immediately. The typewriter on my wife's desk has 45 keys. Imagine that I strike them at random, to produce a line of 72 characters. The chances of obtaining this particular line (or any other that might emerge) are less than 1 in 10^{83}, a number on floor 83 of our tower. It is larger than the number of atoms in the universe. Yet I have tried only once, and there it is! Why not then just ascribe the origin of life to such a lucky, improbable event, end this book, and turn our attention to other questions?

If we did so, we would be running from home plate toward third base once again. If we want to apply science in a consistent way to make sense of the world, then we should fall back on improbable explanations only after we have exhausted the more likely ones.

Suppose, for example, I saw somebody walking across my community swimming pool, with his feet just touching the top of the water. What conclusion should I draw? A very small but finite possibility exists that the water molecules in some part of the pool, which normally shove about in all directions, will at a given moment simultaneously all thrust upward. The area of this thrust might exactly match the size of the walker's foot and occur in just the place where he steps, to support him. Additional portions of the pool might, coincidentally, act in the same way to support his subsequent steps, until he had finished his stroll.

With some effort, I could probably estimate the possibility of this event. I suspect that it would be even less likely than the spontaneous generation of a bacterium. My first inclination, though, if I witnessed the event, would not be to say, "Oh, what a lucky man!" but to look for some trick, or consider how much wine I'd had to drink.

Many of the events described in religion or mythology and considered to be miracles could also be accommodated within the framework of science as extreme improbabilities. But if we were to prefer such an explanation when a more probable one may be available, then we are essentially moving out of science and into a religious position.

Some future day may yet arrive when all reasonable chemical experiments run to discover a probable origin for life have failed unequivocally. Further, new geological evidence may indicate a sudden appearance of life on the earth. Finally, we may have explored the universe and found no trace of life, or processes leading to life, elsewhere. In such a case, some scientists might choose to turn to religion for an answer. Others, however, myself included, would attempt to sort out the surviving less probable scientific explanations in the hope of selecting one that was still more likely than the remainder.

We are far from that state now. Many non-miraculous possibilities are yet open, and we will look at them shortly. But first let us pause to consider one final maneuver.

There is one way in which any event, however improbable, can be made probable. One need only select a model for the universe which postulates that it is infinite. Physicist Michael Hart has done this and written: "In an infinite universe, any event which has a finite probability—no matter how small—of occurring on a single given planet must inevitably occur on some planet."

Thus everything happens somewhere. Our own planet is obviously one place where life began.

Of course, this explanation can be used to legitimize any event. The earth may have been a jumbled heap of chemicals last night. Suddenly, by a random fluctuation, we, our memories, our possessions, and our civilization were created. This event also should occur once in an infinite universe. Perhaps this was the place.

The above argument will not hold if evidence indicates that the universe is definitely finite in its extent. Even without such evidence, the argument is not a useful one. It cannot be refuted, and nothing can be done with it. We had better move on, and look for more satisfying alternatives.

If we reject the idea that life began with the spontaneous generation of a bacterium, or an organism of comparable complexity, then we must presume that some much simpler entity was the first living thing. We then face a formidable question: What was the nature of that entity?

six

The Chicken or the Egg

Imagine that you are the captain of a small sailboat that is sinking slowly in a storm. You must lighten it if it is to stay afloat. Unfortunately, everything that obviously can be spared has already been thrown overboard. What should you sacrifice now: the sail, the food supplies, the radio, the signaling equipment, or perhaps one of the passengers? It is a difficult choice.

A similar dilemma faces the biochemist who considers the origin of life. As we have seen, the simplest known organisms are far too complex to form spontaneously. The hypothetical common ancestor, an organism containing the features shared by living cells today, would also be complex. The first organism was a much simpler one.

What, then, should be sacrificed to strip down the common ancestor into the original organism: the membrane, the energy-generating system, the genetic system, or the vital catalysts? Understandably, controversy exists over this question. It is agreed that one thing must be kept, however. Just as the captain must preserve the hull of his ship, the biochemist must preserve some mechanism in his organism that will permit it to evolve and generate more complex life.

Most biochemists are willing to part with the energy-generating system and to rely upon the benevolence of the prebiotic soup. This soup is called upon to perform the functions of a modern mammalian mother. It must not only assemble a living organism within its body, but it must also continue to nourish it after birth. The chemicals in the soup will furnish the meals for the first organisms, supplying both energy and the substances needed for further growth.

Most biochemists are also willing to forgo the lipid membrane, or to make its acquisition a minor feature in the development of life. If we ignore the protein gateways, then the membrane simply becomes a partition to separate the living cell from its environment. Partitions can be formed in many ways, and need not have complex structures.

Carl Woese, you may recall, suggested compartments made of droplets in a cloud. A foam of bubbles or the interior of a mineral may also provide natural compartments. Certain classes of organic compounds, generally of high molecular weight, may also segregate themselves from a water solution to form tiny droplets. Various types of compounds, not just lipids, can show this behavior. These structures have been called coacervates, and have been extensively studied by Alexander Oparin and others. In a later chapter we will encounter another type of primitive compartment: microspheres made of protein. The formation of compartments is not a difficult business, and this process probably was not the critical one in the origin of life.

When lipids and carbohydrates are thrown overboard, we are then left with proteins and nucleic acids as candidates for the ingredients of the first organism. Some more cautious thinkers would like to retain both of them, but then the boat would surely sink. Both are complex types of molecules, which need to be of considerable size to function properly. We shall see that it is difficult to account for the appearance of either of these molecules by spontaneous generation on the early earth. If both are needed, then we go down in a sea of improbability.

Most workers in the field are willing to face the painful choice. As stated in A. L. Lehninger's biochemistry text, it is: "Which had primacy in the origin of life, proteins or nucleic acids?"

The nucleic acids are, of course, the hereditary material. They contain the blueprint for the organism which is passed from parent to daughter cells. DNA duplicates during replication, to provide a blueprint copy for each daughter. The design of DNA, with its two complementary chains, makes this event possible.

DNA cannot replicate alone, however. It requires the aid of proteins in this process. Further, neither DNA nor the other nucleic acid, RNA, has much catalytic ability. Unlike proteins, they cannot make things happen. Francis Crick summarized it well in *Life Itself*: "RNA and DNA are the dumb blondes of the biomolecular world, fit mainly for reproduction (with a little help from proteins) but of little use for much of the really demanding work."

Any hint that DNA and RNA can do some work is received eagerly by those who favor the primacy of nucleic acids. Late in 1982, for example, Colorado State University chemist Thomas R. Cech and co-workers reported that certain RNA molecules could reorganize themselves. They could rearrange their connections so that certain sections were expelled and others were rejoined. Enzymes could speed up these processes manyfold, but they took place more slowly anyway, even when no enzymes were present.

Science magazine reported the news under the headline "RNA Can Be a Catalyst," and suggested that it had significance for the origin of life. This announcement was premature, as the word "catalyst" has a different meaning. It describes a substance that changes other molecules, while it remains unchanged. Subsequently, other workers showed that one RNA molecule can also aid in the rearrangement, or splicing, of another, in true catalytic fashion.

The effects shown thus far testify to the versatility of RNA as a genetic material, but do not demonstrate the control of other kinds of molecules that would have been valuable in the early days of life. They may have come into play later in evolution, when the partnership of DNA and RNA was first established. As we have seen, the DNA of higher organisms has extra messages ("commercial breaks") that are passed on to RNA, but must be removed before the information is used to construct proteins. The ability of RNA molecules to splice one another without outside help shows how fit they are for this particular role, but tells us little about whether nucleic acids or proteins had primacy in the origin of life.

Proteins can make things happen effectively in the cell. Alas, they lack another capacity. We know of no mechanism by which they can replicate themselves. Like mules, they can work, but are sterile. If a cell were deprived of its DNA, it would function for a time. Cilia would wave, ribosomes would make proteins, and sugars would be converted to simpler substances, releasing energy. After a time, however, everything would grind to a halt. The cell would die, leaving no offspring.

Genes and enzymes are linked together in a living cell—two interlocked systems, each supporting the other. It is difficult to see how either could manage alone. Yet if we are to avoid invoking either a Creator or a very large improbability, we must accept that one occurred before the other in the origin of life. But which one was it? We are left with the ancient riddle: Which came first, the chicken or the egg?

In its biochemical form, protein versus nucleic acid, the question is a new one, dating back no further than Watson and Crick and our knowledge of the structure and function of the gene. In its essence, however, the question is much older, and has provoked passion and acrimony that extend beyond the boundaries of science. In an earlier, broader form, the question asked whether the gene or protoplasm had primacy, not only in the origin but also in the development of life. Ultimately, it can be widened further to question whether heredity or environment is more potent in shaping living beings.

We will enter this arena by considering an article published in 1966 by Nobel laureate H. J. Muller (1890-1967) in the *American Naturalist*, which summarized his views on the origin of life. Muller was an American scientist who had discovered that X rays can produce mutations. He was among the first to warn the public of the adverse health effects of radiation, and was also an advocate of human improvement through voluntary eugenics. He was one of the founders of modern genetics.

Not surprisingly, Muller was the foremost exponent of the primacy of the genetic material in the origin of life. He had suggested this idea in the late 1920s, adapting it from an earlier theory of L. T. Troland. The Troland theory held that enzymes and genes were the same substance (this was long before Watson and Crick) and that this substance, catalyzing its own reproduction, was the master chemical of life. Muller recognized that the functions might be separate, and attached more importance to the gene. We will quote from his 1966 article directly:

> It is the specific sequences in the DNA which determine those in the proteins and *changes* in the former result in corresponding changes in the latter, whereas the reverse relation does not hold, any more than, in general, *other* acquired characteristics are inherited. This circumstance clearly gives the gene material primacy. . . . The "stripped down" definition of a living thing offered here may be paraphrased: *that which possesses the potentiality of evolving by natural selection.* . . . The gene material also, of natural materials, possesses these faculties and it is therefore legitimate to call it living material, the present-day representative of the first life. . . . Primitive conditions afforded it enough means of exercising them to allow it to evolve protoplasm that served it. . . . Thus the gene material itself has the properties of life.

Muller's views do not lack advocates today, among them the astronomer Carl Sagan. Sagan was an undergraduate at the University of Chicago in the early 1950s and spent one summer in Muller's laboratory in Indiana. Subsequently, as a graduate student, Sagan published an article expressing views similar to Muller's:

> The design of the organism is merely to provide for gene multiplication and survival. . . . Now this picture we have been drawing of the protoDNA molecule, associated with protein, is certainly strongly suggestive of a primitive free-living naked gene situated in a dilute medium of organic matter. . . . There was no protoplasm *per se* for the naked gene to act upon. . . . In time the naked gene found it of greater adaptive value to control the environment by becoming no longer naked.

Sagan has continued to advocate this position during his outstanding career in astronomy and science writing. In his book and television series *Cosmos*, the origin of life was equated with the formation of the first self-copying molecule: "the earliest ancestor of deoxyribonucleic acid, DNA, the master molecule of life on Earth."

The nomination of a nucleic acid for the distinction of earliest living entity is in accord with other developments of the last thirty years that have made nucleic acids the most celebrated substances in science, and the darling of the media. The escapades of DNA extend far beyond science and into industry, politics, and ethics.

For example, we hear almost daily of the achievements of Recombinant DNA. Techniques have been developed by which portions of the DNA of one species can be inserted into the DNA of another one, and to function there. Thus bacteria have accepted genes for the production of the amino acid chains of the human hormone insulin. These genes were prepared not in human cells, but in a laboratory. The altered bacteria have been put to work, producing insulin on an industrial scale. In 1982 this product was approved for the market by the U.S. Food and Drug Administration. Many other products will follow.

As the development of these techniques progressed, public fears were aroused about their hazardous potential. For example, there was concern that an altered bacterium containing a cancer-causing gene might escape and produce an epidemic. A temporary moratorium on certain experiments was declared, until effective safety measures could be devised.

With experience and the passage of time, such fears have quieted. The calm may soon be shattered by new developments, however. It is now possible to prepare novel DNA sequences by laboratory synthesis ("designer genes"). Ultimately, proposals will be made to redesign the genes within us, and a new wave of controversy will follow. I write these words immediately after the release of a statement by a group of clergy opposing the use of these methods to alter human heredity. The subject, of course, is one worthy of controversy—the biological future of the human race.

DNA can migrate under natural circumstances as well as artificial ones. Sections of DNA that move from one location to another have been called "jumping genes." Migrations of genetic material between nucleus, mitochondria, and chloroplasts have led to the term "promiscuous DNA." The behavior of this mischievous molecule in other circumstances has earned it additional titles: skeletal DNA, parasitic DNA, dead DNA, ignorant DNA, and selfish DNA.

The last term was applied by Francis Crick and Leslie Orgel to certain DNA sequences that have no function of their own, but have intruded among working sequences in a way that makes it too expensive (in energy) for the cell to remove them. They are perpetuated as a molecular parasite on the useful DNA.

The word "selfish" has also been applied to DNA in a broader sense by Richard Dawkins in *The Selfish Gene*. He assigns to DNA the central role in the development of life. The remainder functions merely as a means to provide for the survival and propagation of DNA. In this view, the body of an elephant is merely an elaborate machine devised by elephant DNA to ensure its own perpetuation.

The rise of nucleic acids to their current state of eminence and power represents an authentic Horatio Alger, rags-to-riches story at the molecular level. Their beginnings were humble indeed.

A nucleic acid was first isolated in the laboratory of a Swiss chemist, Friedrich Miescher, in 1869. The source was rather unappetizing: pus cells from surgical bandages. The discovery was greeted with skepticism. Miescher's mentor, E. F. Hoppe-Seyler, insisted on repeating the result himself before it could be sent for publication.

Obscurity was Miescher's reward for his achievement. This lasted for his lifetime and well after his death. Upon the occasion of the centennial of Miescher's discovery, in 1969, the biochemist Erwin Chargaff noted:

> I should like to start this essay with one of the quiet in the land, with Friedrich Miescher, who a hundred years ago, in 1869, discovered the nucleic acids, somehow between Tübingen and Basel. As was to be expected, nobody paid any attention to this discovery at the time. The giant publicity machines, which today ac-

company even the smallest move on the chessboard of nature with enormous fanfares, were not yet in place. Seventy-five years had to pass before the importance of Miescher's discovery began to be appreciated. Miescher himself—and this appears clearly from his correspondence and from the tone of his compact papers—was well aware of the importance of his observations. They failed, however, to make much impression on his time; and how little echo there was can perhaps be deduced from the fact that even today the best history of the natural sciences, in the volume devoted to the 19th century and published in 1961, mentions the name of Darwin 31 times, that of Huxley 14 times, but Miescher not at all. There are people who seem to be born in a vanishing cap.

Ironically, Chargaff himself had made a vital but perhaps underappreciated discovery about the composition of DNA, one crucial to the Watson-Crick theory, earlier in his career.

The nucleic acids, like their discoverer, continued in relative obscurity long after 1869. They were known to be present in the cell nucleus, but their function was unclear. Most biochemists felt that if a hereditary chemical did exist, it was likely to be a protein. A small number of heroic chemists nonetheless set themselves the task, over half a century, of determining the structure of nucleic acids.

I term them heroic because the properties of nucleic acids were obnoxious, in comparison with those of simpler organic chemicals available in abundance for structure determination. The nucleic acids would not distill or form crystals or dissolve in convenient solvents such as benzene. Working with them initially required indirect and laborious techniques.

Yet the obsessive persistence of the chemists eventually paid dividends, and the final details of the fundamentals of nucleic acid chemistry were added by Alexander Todd and co-workers at Cambridge University in the 1940s and early 1950s. The time and place were well chosen: the stage was set for Watson and Crick.

Hints had appeared earlier of the forthcoming importance of DNA. In 1944, Oswald Avery with his colleagues Colin McLeod and Maclyn McCarty had published an unexpected result. The heredity of certain bacteria could be altered by treating them with "transforming principle," a DNA preparation from related bacteria. This experiment made little immediate impact but ultimately had a profound one. By that time viruses had been recognized as infective entities made only of nucleic acid and protein. Alfred N. Hershey and Martha Chase reported in 1952 that the DNA, and not the protein, carried the information of heredity.

The double helix structure was published in 1953. After another decade the genetic code was deciphered. The era of Recombinant DNA began in the early 1970s. Since that time, the many other details of the function of nucleic acids and proteins have come out. We understand much of the chemical basis of heredity.

These remarkable developments were not everywhere welcomed, however. They received a notably unpleasant reception in the Soviet Union. With this theme we will return to the gene-versus-protoplasm conflict and the origin of life. Once again, our entry point will be the 1966 *American Naturalist* article by H. J. Muller, in which he wrote of the protoplasm-has-primacy view:

> It is a curious anachronism . . . that even today some of the most eminent biochemists and biologists, doing very valuable work in their respective fields, still adhere to this view and its corollary concerning life's origin. Unfortunately, it became much publicized and elaborated, beginning in the 1930's by the Lysenkoist Oparin in his book *The Origin of Life* (1938 et seq.) as a part of the attempt to down-rate the significance of genetics. His part of that attempt was most subtly carried out.

We have encountered Oparin earlier in this book (and shall meet him again), but not Lysenko. Further, the depth of feeling in the above statement transcends purely scientific and rational disagreements. To penetrate this matter further, we must consider the life of H. J. Muller in greater detail.

He was a New Yorker by birth, born in 1890, and he received both his undergraduate and graduate degrees from Columbia. While he was there, he became involved in the research conducted by the group headed by Thomas Hunt Morgan. They worked with the Drosophila fruit fly, which proved an ideal vehicle to explore the basic principles of genetics. Pioneering work in the mechanism of heredity had been done by an Austrian monk, Gregor Mendel, forty years earlier, and then forgotten. After the rediscovery of Mendel, the Morgan group performed the outstanding studies that identified the role of genes and chromosomes.

Muller himself performed his most notable work while on the faculty of the University of Texas, from 1920 to 1932. He discovered the mutational effects of X rays in that period. He was elected to the National Academy of Sciences in 1931. He suffered, however, from conflicts with his colleagues, a failing marriage, and a growing dissatisfaction with social conditions in the United States, particularly during the Depression. His strong socialist views ultimately caused him to leave this country.

Muller moved to the Kaiser Wilhelm Institute in Berlin in 1932, only to see Hitler rise to power. He then received an invitation from the noted Soviet geneticist Nikolay I. Vavilov to assume the directorship of a genetics laboratory in the USSR. He accepted, and eventually established his institute in Moscow.

Whatever joy he may have felt at the union of his research interests with his political convictions was short-lived. For this was the period when Trofim D. Lysenko rose to power in Soviet biology.

Lysenko was essentially an agricultural reformer who advocated the ideas of an uneducated breeder of fruit trees, Ivan V. Michurin. To be brief, Lysenko believed in the inheritance of acquired characteristics and denied the importance or even the existence of genes and chromosomes as units of heredity. As quoted in the account of Soviet dissident Zhores A. Medvedev, Lysenko said: "The hereditary basis does not lie in some special self-reproducing substance. The hereditary base is the cell, which develops and becomes an organism. In this cell different organelles have different significance, but there is not a single bit that is not subject to evolutionary development."

This Lysenko version of heredity, which was termed Michurinist, was contrasted to the Mendel-Morganist view of the gene, which was considered formalistic, bourgeois, and metaphysical science. These conclusions flowed not from a careful weighing of experimental evidence, but rather from Lysenko's perception of the ideological needs of the state.

Michurinist ideas found favor in the USSR because they harmonized well with the prevalent philosophic theory of communism—dialectical materialism. The primary thrust of this philosophy deals with the development of societies, historical forces, class struggles, and other matters which need not concern us here. However, Friedrich Engels, one of the two nineteenth-century germinal thinkers of the socialist movement (Karl Marx was the other), had taken an interest in the development of life as well as the evolution of societies. Engels had written: "Life is the mode of existence of albuminous substances and this mode of existence essentially consists in the constant self-renewal of the chemical constituents of these substances by nutrition and excretion." The term "albuminous," in its most general sense, merely refers to water-soluble proteins. One prominent form is ovalbumin, a substance in egg white which serves as a nutrient for the developing chick embryo.

In any event, Engels felt that life and humanity had resulted from a continuous evolution of matter, the origin of life being only a rung in the long ladder of development. At a much higher level, the same evolutionary process led societies toward socialism.

One plausible extension of these concepts was the idea that the environment shaped heredity. The connection was expressed clearly by Muller's biographer, E. A. Carlson. The socialist state had made dramatic changes in literacy, employment, and other social areas. Why should it not also be able to affect hereditary afflictions, such as mental retardation and certain diseases? It seemed reasonable to presume that a better form of man would be produced by an improved environment.

Thus the inevitable development of life became a theme of Marxist philosophy. Rejected equally in the dialectical materialist view were idealism (the name for the school of philosophy that emphasizes the role of spiritual values in existence) and mechanism. The latter term was applied to any belief in spontaneous generation, the role of chance in the origin and development of life, or the idea that the higher properties of matter could by deduced directly from the basic laws of physics and chemistry. Dialectical materialism held that new laws, biological, social, or whatever, came into play as matter reached higher levels of development.

The main interest of the Lysenko group was not in biochemical matters, however, but simply in improved farming methods. They believed that soaking of seeds ("vernalization") could convert winter wheat into spring wheat. By similar methods, other plant species could be interconverted. They hoped that their new biology could revolutionize agriculture. Ultimately, they failed because their methods simply did not work. In the attempt, they crushed Soviet genetics for a generation in their campaign that produced "thirty-five years of brutal irrationality" (according to Soviet scholar David Joravsky).

H. J. Muller found that he had moved his laboratory directly into the path of this tidal wave of stupidity. He and his colleagues were advocates of the gene theory. The name of his former mentor, Morgan, had become a synonym for bourgeois decadence. In 1934, before the new ideology was fully formed, Muller made an attempt (which he subsequently regretted) to relate chromosome theory to dialectical materialism. This did not work.

Muller personally considered Lysenko to be a fraud and a gangster. He defended gene theory and opposed Lysenko's views at a Soviet conference in 1936. He pointed out that Lysenko's views were ultimately derived from those of a French philosopher, Jean-Baptiste de Lamarck. The notion of the inheritance of acquired characteristics, called Lamarckism, had been discredited by a number of experimental studies. For example, August Weismann, a nineteenth-century German biologist, had amputated the tails of hundreds of mice over five generations, only to find that all the progeny grew normal, not even shortened, tails.

In addition to pointing out the relation of Lysenko's ideas to those of Lamarck, Muller claimed that Lysenko's views on heredity were a logical basis for racism and fascism. This drew applause from academic delegates but naturally did not bring the approval of those attacked. As time passed, Muller and his co-workers were subjected to increasing harassment. Ultimately, Muller had to leave the Soviet Union. He volunteered for service in the Spanish Civil War, and returned to Moscow only to pack his belongings. He was an academic migrant for a time, until he eventually was able to secure a job at Indiana University in 1945. The next year he was awarded the Nobel Prize.

Muller's Soviet sponsor and friend, Vavilov, suffered a less kind fate. He became a leader of the group opposing Lysenko. In 1940 he was arrested, tried, and imprisoned. According to Medvedev, Vavilov was maltreated in prison and died in Siberia.

Lysenko did not come to the full height of his power until 1948, following a meeting at which five prominent geneticists who had opposed him recanted and changed their views. In the words of Medvedev, Lysenko's followers then "greedily grabbed ranks, posts, scientific degrees, honorary titles, prizes, salaries, medals, orders, honorifics, honoraria, apartments, summer houses, personal cars. They did not just await bounties from nature."

Muller had remained silent since leaving the USSR, lest he endanger former colleagues and associates who had remained there. But at this point, he resigned from the Soviet Academy of Sciences and denounced Lysenkoism. The Academy noted without regret the departure of "its former member who betrayed the interests of real science and openly joined the camp of the enemies of progress and science, of peace and democracy."

As we have seen, the period following World War II was the time of ascendancy of molecular biology. Lysenko's fortunes rose and fell with political developments, however, rather than scientific ones. Joseph Stalin's support had been a source of his power. Lysenko came under attack after Stalin's death in 1953 and he was forced to resign certain offices in 1955. He then found new strength and support with the ascendancy of Nikita Khrushchev, and lost his power only after the ouster of Khrushchev in 1964.

During the last period of power of the Lysenko group, the rhetoric used in defense of their scientific views took on a tone more usually associated with Soviet political pronouncements. For choice examples of them we are again indebted to Zhores Medvedev.

We can begin with a 1951 quote from Olga Lepeshinskaia, a cell biologist whose work on spontaneous generation we noted briefly. We shall soon encounter her again.

In our country there are no longer mutually hostile classes. Yet the struggle of the idealists against dialectic materialists, depending on whose interests are being defended, still has the nature of class war. And, in fact, the followers of Virchow, Weismann, Mendel, and Morgan, talking of the immutability of the gene and denying the effect of the environment, are preachers of pseudo-scientific tidings of bourgeois eugenicists and of various distortions in genetics, which provided the base for the racist theory of fascism in capitalistic countries. World War II was unleashed by imperialist forces whose arsenal also included racism.

We already have met Weismann, Mendel, and Morgan. Rudolf Virchow was a nineteenth-century pathologist who studied disease at the cellular level. I am not certain which particular achievement moved him to the head of the above list.

The advances in genetics that followed the Watson-Crick theory did not alter the views coming from Lysenko and his followers. We will quote from an article from N. M. Sisakhan in 1954, after the publications by Watson and Crick:

In the past, to explain the supermaterialism of living phenomena, vitalism advanced concepts of entelechy or vital force. Its current variety, in the guise of Morganism, resorts to genes, codes, and templates in order not to lose its scientific aspect. But, as we know, changing terminology does not change the substance. And, in substance, entelechy, template molecules, vital force, genoneme are all synonyms. No matter what contrivances Morganists use, they cannot help that their only purpose in juggling the new terminology is to camouflage the idealistic essence of their doctrine and to cover undisguised idealism with a scientific sauce.

In 1962, as the details of the genetic code were being deciphered, the following statement was made in an article by K. Y. Kostrinkova: "The hypothetical connection of the empty abstractions [of the gene theory] with specified substrates—chromosomes, DNA—declared to be the 'material carriers of heredity' does not confer on these abstractions material content, any more than superstitious deification of objects makes the superstition materialistic." Lysenko himself in 1963 continued to deny the existence of a hereditary substance or the role of DNA in inheritance.

These typical declarations were not accompanied by a scientific critique of the theories they opposed. There was no analysis in detail of particular experiments, no reference to flaws in methodology or logic. Further, they did not perform comparable biochemical studies of their own which led to opposite conclusions. In the words of Medvedev: "The basic activity of Lysenko's followers in the theoretical field both then and now consists of misinformation and criticism, and, as before, they consider their main service to be the struggle against their opponents." Further, the nature of their criticism was to label their opponents' views as superstition, or in other words, the wrong religion. They, by contrast, had derived their views on heredity from the principles of dialectical materialism, which flowed from the ideas of Engels and Marx. In short, they had the right religion.

There is a strong parallel here to the Creationist controversy, which we shall consider in a later chapter. In each case, a large and exceedingly well documented body of scientific data, and the conclusions derived therefrom, were dismissed as religion. The opponents of these conclusions had little valid data, or none at all, as their views were essentially derived from religion or myth. Yet they preferred to reserve the term "science" for themselves. One important difference between the two cases is that the Lysenkoites were supported by the full power of a totalitarian state.

After the fall of Lysenko, Soviet genetics gradually recovered and returned to the modern world, though Lysenko himself retained his titles and was free to advocate his views, until his death in 1976. After 1964, however, Mendel's name could be mentioned with respect once again. By 1969, geneticist N. P. Dubinin was describing mutations in terms of dialectical principles. Thus, the methodology could adapt to changing circumstances.

In an article in *Nature* in 1983, a vice-president of the Soviet Academy of Sciences gave a confident appraisal of Soviet biotechnology. Scientists in the USSR had been aware of the potential of the new techniques for a decade, and were using them to produce insulin and human growth hormone in altered bacteria. A chief objective was the manipulation of plant genes in order to increase food production.

It is ironic that the very goals of the Lysenko movement, interconversion of species and improved agricultural production, could be approached best by the methods of the field they despised—and one that they held to contradict socialist dogma. These methods may ultimately even be applicable to a higher socialist ideal: the improvement of humanity itself.

We have wandered astray, however, from the subject of the origin of life, and it is time to return. In particular, we wish to consider the career of Alexander I. Oparin (1894–1980), which spanned all of these difficult times in the Soviet Union.

Oparin was a key contributor to the modern paradigm of the origin of life. We have discussed some of his ideas in terms of the Oparin-Haldane hypothesis and the role of coacervates. In an obituary published in *Transactions in Biological Sciences*, he was called "the acknowledged leader of the international community of scientists studying the origins of life." He was the first president of the International Society for the Study of the Origin of Life. A medal was created in his honor by the society after his death. He was honored in his own country as well. For many years he was director of the Institute of Biochemistry of the Academy of Sciences of the USSR. He was awarded the Order of Lenin, made a Hero of Socialist Labor, and given other distinctions. Though he spoke no English, he made a good impression on his visits abroad. The obituary mentioned above paid tribute to his friendliness toward foreign colleagues and to his remarkable hospitality.

Oparin's views on the origin of life were first delivered in an address to the Botanical Society of Moscow in the spring of 1922, and were published in 1924. They received little notice at the time. J. B. S. Haldane was unaware of Oparin's work, and published similar ideas in 1929. In a 1963 meeting, Haldane graciously conceded priority to Oparin: "I have little doubt that Professor Oparin has the priority over me. I am ashamed that I haven't read his early work so that I don't know . . . there was

precious little in my small article which was not to be found in
his books. . . . The question of priority doesn't arise. The question of plagiarism might."

Oparin published a book in 1936 which made a much fuller
statement of his theories. This book was translated into English
in 1938 and brought him an international reputation. There were
significant differences between this work and his earlier one,
however. Both featured the reducing nature of the early earth,
which permitted the synthesis by ordinary chemical reactions of
a sea of organic compounds (Haldane's "hot dilute soup"). Both
versions saw life arising from this soup. The initial organisms
that developed then used the soup as a food supply for a time.
(An earlier prevailing view had been that the first organisms
made their own organic substances.)

How did this remarkable change from soup to living being
take place, however? Oparin, in his original statement, felt that it
came about by random processes: "It is impossible, incredible, to
suppose that in the course of many hundreds or even thousands
of years during which the terrestrial globe existed, the conditions
did not arise 'by chance' somewhere which would lead to the
formation of a gel in a colloid solution." This last structure was
associated with the first primitive living system by Oparin. It is
the same structure he later identified by the word "coacervate."
If we expand the time scale somewhat, we have essentially the
position later stated by George Wald: spontaneous generation.

In the 1936 book, and in the later works, Oparin stressed a different mechanism: gradual and inevitable chemical evolution.
This viewpoint was fully in accord with emerging Marxist views
on heredity. According to the account of David Joravsky, in
Oparin's 1924 work there had been "not a breath of Marxism in
this pamphlet, conscious or unconscious." In the 1920s, Marxist
biologists did not see the origin of life as a topic that distinguished them from their non-Marxist counterparts.

But, according to Joravsky again, "in the 1930's, when con-
fession of Marxism was imposed on the Soviet intelligentsia,
Oparin became one of the most active confessors. He began by
claiming that Engels was one of the originators of his approach to
the origin of life. . . . He altered his speculations on the origin of

life to suit the Lysenkoite creed, suppressing consideration of the origin of genetic systems."

In fairness, I must point out that Oparin's later views were probably derived not only from political expediency but also from conviction, as he came to them before it was fully necessary to do so and held them to his death, long after the fall of Lysenko. What were these ideas? Let us quote him:

> According to the dialectic materialist view, matter is in constant motion and proceeds through a series of stages of development. In the course of this progress there arise ever newer, more complicated and more highly evolved forms having new properties which were not previously present. . . .
>
> Now even the biological laws have been driven from the foreground and the laws of development of human society have begun to play the leading part in further progress.

Oparin, after 1936, denied spontaneous generation, stating that it was inconceivable that living entities "could appear in a very short time, before our eyes, so to speak, from unorganized solutions of organic substances." On these grounds, he rejected the concept of the naked gene—the sudden appearance of a molecule well adapted to its function. He denied that life could be inherent in an individual nucleic acid or protein molecule while the rest of the protoplasm was merely a lifeless medium. He often compared such views to those of the ancient Greek philosopher Empedocles, who had felt that living things had appeared by the independent development of separate organs—arms, eyes, ears, and so on—which then came together.

Oparin's views were certainly adequate to ensure his survival through the Lysenko period. His services to this cause, however, exceeded this minimum necessity. In Joravsky's words: "Oparin was the only really distinguished biologist who gave really strong support to Lysenkoism." Medvedev is quite critical of Oparin's role, stating, for example, that he went out of his way to praise Stalin as "the inspiration of progressive biology": "According to Oparin, Stalin, long before Lysenko, asserted that acquired characteristics are inherited, and that it was precisely these 'strokes of Stalin's genius' that inspired the Michurinists in their fight against neo-Darwinism as an idealistic perversion of biology."

During the period 1948–1955, Oparin served as academic secretary (that is, chief) of the Biological Section of the Academy of Sciences and had a role in filling important vacancies. Medvedev comments on the case of D. A Sabinen, an important plant physiologist who had fallen out of favor. He was dismissed, but after years of effort found a new position. "But Oparin, then heading the AS Biological Section, and who fawned on Lysenko in every way, flatly refused to approve Sabinen's appointment and he once more became an outcast." Eventually the poor man shot himself.

In 1950, Oparin had to join with Lysenko to support the award of a Stalin Prize to Olga Lepeshinskaia. Loren Graham, a historian of Soviet science, describes her as "a mediocre biologist of impressive political stature." She had been a Communist Party member since the time of its founding, and had a personal association with Lenin and other political leaders. We have already seen an example of her prose style.

Her scientific work included claims that she could prepare living cells from noncellular nutrient media, in as short a time as twenty-four hours. One such preparation included the use of albumin, from egg white (Engels' words were apparently taken quite literally). Medvedev states that she "pronounced the great Louis Pasteur a reactionary and idealist." Lepeshinskaia was able to refute him by obtaining spontaneous generation in broths of hay. Her other achievements included the discovery that soda baths constituted an important anti-aging measure.

Oparin supported her award and praised her great service to science. Graham believes that Oparin acted in response to political pressure, as her views certainly contradicted his own. Afterward, he gradually backed away from this position and returned to flat opposition to spontaneous generation. For this, he drew criticism from Lepeshinskaia and her supporters.

A temporary rebellion aganst Lysenkoism took place in 1955, during the transient thaw and liberalization that followed Stalin's death. At that time, according to Medvedev, three hundred Soviet scientists signed a petition requesting the removal of Lysenko and Oparin from their positions in the Academy of

Sciences. These requests were granted. In the obituary of Oparin in *Transactions in Biological Sciences,* his entire involvement with this job is summarized briefly: "Oparin was also Secretary of the Biology Division of the Academy during an unhappy period, 1948–1955."

It was in this climate that the First International Symposium on the Origin of Life was held in Moscow, in August 1957. The early 1950s had seen the publication of the Watson-Crick theory and the Miller-Urey experiments. Miller in his paper had acknowledged a debt to the ideas of Oparin. In 1955 the idea of an origin-of-life symposium had been suggested at an assembly of the International Union of Biochemistry. The conference organizers felt that the Soviet Union, "the scientists of which had made a considerable contribution to the solution of the problem of the origin of life," was an appropriate place for the meeting. The time was also right, as "a certain turning point" had taken place in the field.

The conference itself provided a forum for the expression of opposing views on how life arose. Oparin presided, and gave an introductory address. H. J. Muller did not attend, but a number of American scientists were present who shared his position on the living gene, among them Norman Horowitz, a biologist from the California Institute of Technology. Nor was Lysenko present, but a number of his supporters were in attendance. Some Western scientists joined with the Soviet scientists in endorsing the gradual evolution point of view. Olga Lepeshinskaia was there. She cited her own studies and quoted the definition of Friedrich Engels on the nature of life.

Historian John Farley summarized this first grand conference on the origin of life in the following way: "Behind the seemingly innocuous questions being posed, there lay deep ideological and political differences which loomed large in the cold war of the 1950's." These questions were undoubtedly present, but not necessarily obvious. I myself was a graduate student at Harvard at the time, and not involved in any way. I recently asked my friend Bea Singer, an attendee, for her impressions. Travel to the Soviet Union was novel at that time. Bea mostly remembered only the travel arrangements, and nothing of a political confrontation.

The conference did, in any event, inaugurate a continuing series of international meetings on the origin of life. The Second International Conference was held in Wakulla Springs, Florida, in October 1963. A feature was the first meeting of Oparin with J. B. S. Haldane, the co-founder of the central paradigm.

Haldane differed from Oparin in that the origin of life had not been a central concern of his scientific career. His reputation had been earned as a mathematical biologist, geneticist, and physiologist. He did share with Oparin a devotion to communism. Haldane had adopted Marxist views in the 1930s and served for several years as editor of the *Daily Worker* of London. While Haldane supported the Communist Party line on most matters, he ran into difficulty on Lysenkoism, particularly after the events of 1948. He apparently disliked the treatment given to Lysenko's opponents, but was in doubt over the validity of the scientific ideas, feeling that there might be something to them. His biographer, Ronald Clark, reports that Haldane asked Lysenko for experimental details. When he did not receive them he broke with the party, writing in 1949: "I am a Mendelist-Morganist."

In the latter part of his life (he died in 1964, one year after the Florida conference) Haldane also broke with his native country, England. He moved to India in 1957 and took up Indian citizenship. Dissent was apparently part of his nature. The "hot dilute soup" theory had been novel when he proposed it. When it gained acceptance, years later, his mistrust of orthodoxy made him doubt whether it could be correct.

Haldane and Oparin, the two original chefs of the prebiotic soup, disagreed on the manner in which life arose in it. Haldane was the only Marxist who favored spontaneous generation. At the 1963 conference both reiterated their points of view, with Haldane stating that "the initial organism may have consisted of one so-called gene of RNA specifying just one enzyme."

Despite this, the two apparently got along well. Haldane was chosen to introduce Oparin and stated: "I suppose that Oparin and I may be regarded as ancient monuments in this branch of science but there is a very considerable difference, that whereas I know nothing serious about it, Dr. Oparin has devoted his life to this subject."

If H. J. Muller had been present, he might have shown less

deference. Muller missed the conference due to serious illness, however. Subsequently, he read the proceedings and noted that only a handful of participants, including Haldane, had taken his position, while many others espoused the protoplasm-primacy view of Oparin.

This shift of opinion, from spontaneous generation of a naked gene to the gradualism of Oparin, is a central theme of the book *Spontaneous Generation from Descartes to Oparin* by historian John Farley. As a mark of this shift Farley notes the publication of a widely used text by John Keosian, an American biochemist "with no known Marxist affiliations." In this text, Keosian writes: "From the materialist view, the origin of life was no remote accident; it was the result of matter evolving to higher and higher levels through inexorable working out at each level of its inherent potentialities to arrive at the next level." Farley himself concluded in 1974:

> Today the majority of biologists and biochemists seem committed to the evolutionary viewpoint of Oparin . . . . Life did not arrive by a spontaneous generation. That is to say, that a functional living entity, whether that be a mouse, maggot, bacterium, virus or "living molecule," did not make an all-at-once appearance from material with no lifelike qualities. Life emerged slowly as part of a long developmental process, all stages of which were highly probable at the time they occurred.

Oparin himself, and his ideas, clearly were survivors. Oparin was able to pass through and beyond the Lysenko period without severe difficulties. He was able to appear in a dual role, as a Lysenko partisan at home and as a benign theorist on the origin of life in the Western world. By 1964 he had moved wisely to a position of neutrality on a question concerning a Lysenko appointee. Throughout his career he was able to maintain a secure platform for the dissemination of his scientific views.

His careful attitude on these matters comes through in an account of a 1978 Moscow interview by the journalist Harold T. P. Hayes. The interview was conducted in the presence of a deputy director of the Academy and a translator. Hayes was asked to submit questions in writing, and Oparin chose to answer only certain questions from the list. At the end of the interview, brandy, cookies, and candy were served, and a longer response in writing was promised. But only a greeting card arrived, a year later at Christmastime.

Oparin's views are the sole surviving fragment of Lysenkoite biology, derived from dialectical materialism. Unlike the remainder of Lysenkoite thought, they may even have some validity. They certainly, as Farley noted, have gained some ascendancy. Farley, however, hedged his bets. He concluded: "The issue is one which has been abandoned many times before, only to reappear at a later date under a different guise. Whether the final chapter of the spontaneous generation controversy has now been written it is impossible to say."

Oparin, Muller, Haldane, and Lysenko have all passed on. The political heat of the question has also passed and become part of history. The scientific aspects remain. The Skeptic, in fact, has grown very restless during this discourse. He points out that the political complications are irrelevant to the scientific answer. When Oparin stated, for example, that "only dialectical materialism has found the correct routes to the origin of life," he was contributing dogma rather than experiment. Science works neither by pronouncement nor consensus, but rather by experiment.

In fact, the naked gene theme, with its corollary of spontaneous generation, is very much alive today. It has drawn new strength from recent experimental studies and mathematical treatments, and we will turn our attention to these in our next chapter (which is unlikely to be the final one in its history).

seven

The Random Replicator

Origin-of-life scientists disagree on many topics. One vital area of conflict lies between those who believe in chemical evolution and those who propose the naked gene, whom we shall call naked genies. As we have seen, this quarrel has transcended science and spilled over into politics and philosophy. In this chapter we shall explore the beliefs of the naked genies in greater depth.

The best-known mechanism that functions to increase the complexity of species is Darwinian natural selection. It has served to guide the development of the first one-celled organisms into the variety of higher beings, including man, that inhabit the earth today. If we accept this current scientific view, we are still left with a mystery: How did the first one-celled creatures arise? They are too complex to form by spontaneous generation, and so must also be the products of evolution from even simpler beings.

The origin of life, according to a naked genie, would coincide with the appearance of the first entity that had the ability to reproduce itself and undergo mutations. Some of these mutations would lead to the creation of descendants more fit for continued survival. These survivors would proliferate and continue the process of evolution by natural selection.

It becomes important, then, to find the simplest possible self-reproducing, or replicating, system, for this would be the first living thing. In this search, viruses are an obvious source of inspiration. They are made of relatively short lengths of nucleic acids, wrapped in protein. When bottled and on a shelf, they appear as an innocuous white powder, hardly distinguishable from sugar or salt. A preparation of tobacco mosaic virus, for example, may sit harmlessly in a jar for months or years. When a portion of this powder is applied to the leaves of a tobacco plant, however, it causes a disease, with the appearance of mottled lesions on the leaves of the plant. In this process, the virus multiplies itself many times.

To appreciate how viruses relate to other living organisms in size and complexity, let us call once more upon our imaginary elevator, COSMEL. We will move to the -6 level, where common bacteria are about our own size and atoms are barely visible. At this level, viruses vary in size from as small as a quarter to as large as our forearm. Some are round, others more cylindrical, while others yet have much more complicated shapes. We will turn our attention to one of the larger viruses, one called T2, which resembles a toy moon lander. It has a hexagonal head, a complex neck shaft, and six spindly jointed legs, all made of proteins. More than fifty different proteins are used to build this structure. Within the head is tucked a length of DNA, which stores the genetic information. T2 is elaborate, as viruses go, and contains more than 100,000 nucleotides on each chain of DNA. Viruses are parasites, and cause diseases in humans ranging from colds to cancer. The one we are observing, however, has no interest in us but chooses bacteria as its target.

The life cycle of a T2 virus proceeds in the following way. It lands with its legs on a bacterial surface and squats to bring the end of its neck shaft in contact with this surface. The T2 DNA is injected through the neck of the virus into the bacterium. This DNA immediately directs the production of RNA molecules and proteins, using bacterial ribosomes, enzymes, and supplies of subunits for the purpose. An enzyme is prepared that destroys the DNA of the bacterium. T2 DNA takes over the cell and converts it into an assembly line for the production of proteins and DNA, to make more T2 virus particles. After a time, the bacterium bursts open and hosts of new viruses are released, to seek fresh victims.

In the T2 life cycle, the nucleic acid is the essential part of the virus, while the protein coat serves for protection and to transport the nucleic acid from one victim to another. The protein functions as a combination overcoat and automobile.

Many viruses are simpler than T2. A number of the smaller ones use RNA, rather than DNA, as their genetic material. RNA molecules are generally much shorter in length than DNA, but they share the same ability to form double helixes, store information, and be replicated.

Of particular interest to us is Qβ, which like T2 is a parasite of bacteria. It has as its genetic material a single strand of RNA, which contains about 4,500 nucleotide units. A nucleic acid need not be a double helix all the time. It must take this form when it is being copied, however, as in replication. Because of the short length of its nucleic acid, Qβ virus can code (hold the instructions for) only a few proteins. It cannot afford an elaborate coat in the style of T2, but must settle for a much humbler one made of a single type of protein stitched together repeatedly.

There exist even smaller RNA molecules capable of reproducing themselves. Some, the viroids, are bits of circular single-stranded RNA with just a few hundred nucleotide units. Viroids are naked, as their stiff rods of RNA come unclothed in any protein coat. We cannot call them genes, however, as their nucleic acid appears not to code for any proteins. Just the same, viroids can replicate within certain plants and cause diseases. They carry names such as potato spindle tuber viroid and cadang-cadang disease viroid. I once saw slides of an entire palm plantation blighted by the disease caused by the latter viroid, and found it awesome to think that the damage had been caused by a replicating agent of such tiny size. On the -6 level of COSMEL, a viroid would be no larger than a finger.

Certain RNA molecules even smaller than viroids are capable

of replication. They do not occur naturally, but are derived from
$Q\beta$ virus in experiments that have demonstrated an analog of
Darwinian evolution in the test tube.

In replication, $Q\beta$ RNA is copied to make more $Q\beta$ RNA. Bacteria contain enzymes that replicate DNA, and others that convert DNA messages into RNA, but no enzymes that copy RNA.
Such enzymes, if they existed, would go about copying transfer
RNA molecules and the RNA in ribosomes, whether these extra
copies were needed or not. The amounts of RNA present in a
bacterium are normally controlled by the DNA of the cell, not by
the direct reproduction of RNA.

Thus $Q\beta$ must supply its own copying enzyme, the replicase,
if it wishes to have progeny. Soon after entering a bacterium, $Q\beta$
RNA acts as a messenger and directs the manufacture of this enzyme, using bacterial ribosomes for the purpose. The replicase
first converts the RNA of $Q\beta$ to a double helix, and then uses this
to make more copies of the original $Q\beta$ RNA strand. Supplies of
raw materials are needed for construction. The appropriate ones
are certain forms of nucleotides which contain a built-in supply
of chemical energy. We will call them "active nucleotides" in this
book. The virus has no active nucleotides of its own; it uses those
of the bacterium. The replicase has another vital ability. It can
distinguish $Q\beta$ RNA from the various bacterial RNA present,
and does not waste its time copying the latter.

The RNA of $Q\beta$ can reproduce in a test tube as well as in a
bacterium. This system was first explored in a notable series of
experiments conducted at the University of Illinois by the biochemist Sol Spiegelman. The replicase is needed, of course, as
well as subunits to construct the RNA and certain salts to keep
the replicase and $Q\beta$ RNA in good condition. Nothing more is
required. When these ingredients are mixed together, the RNA is
replicated until the components in the test tube are exhausted.

If more of these subunits are added, the process can continue
indefinitely, but more room is needed for the progeny. To avoid
transferring the contents of the tube first to a sink, then to a
bathtub, a simple expedient is used. A sample of the original test
tube in transferred after a time to a fresh tube containing additional active nucleotide subunits and replicase, but no new RNA.

By this method, the descendants of the original RNA molecules can be followed for dozens of generations.

Errors inevitably occur when the replicase copies its RNA. When such mutations occur within the normal life cycle of Qβ virus, they may lead to a change in the amino acid sequence of a protein coded by the RNA. If the new protein produced is badly defective, the mutant virus may be unable to survive or replicate. The change will have been extinguished.

Many mutational changes in Qβ RNA may produce only a slight disadvantage. The new protein may be as good, or almost as good, as the original one. Further, because of the way in which genes are stored within the RNA of Qβ, and the nature of the genetic code, some RNA sequence changes may not affect the proteins at all—they can be harmless.

Individual Qβ viruses in nature are usually not identical, because of the continual occurrence of mutation. One RNA may differ from another by the identity of one or two nucleotides, in a sequence of thousands. A group of Qβ virus particles is essentially a crowd of very closely related individuals, with a common average heredity, reflected by the average RNA sequence. No individual virus is likely to deviate very far from this sequence, however, for as mutational changes accumulate, the chances of obtaining one that is lethal increases. Much rarer is the truly beneficial change which by natural selection would come to dominate the population.

The rules of the game change considerably, however, when Qβ RNA is allowed to replicate in a test tube. Its situation can then be compared to that of a wild animal in a zoo. It is protected from danger, and need not hunt for food. Its needs are supplied by the keeper (Spiegelman, in the earlier experiments). It has nothing to do but breed. Under these conditions, mutations in the coat or replicase proteins are harmless. The RNA needs no coat, and the replicase is provided externally. No protein at all can be manufactured, as the RNA has no access to the necessary machinery within a bacterium. Changes in these sequences will not be harmful. A different type of change will be beneficial, however: one that speeds up its replication.

If an RNA molecule can be replicated in ten minutes, say,

rather than the usual twenty minutes, it will have two generations in the time normally needed for one, and four descendants rather than two. Eventually, the progeny of this single molecule will come to dominate the entire mixture. The slower competing ones will gradually be lost in the various dilutions. This mimics Darwinian natural selection: we have the survival of the fittest RNA molecule.

By what means can an RNA molecule speed up its own reproduction? The most obvious method is to shorten itself. Just as we can copy by hand a message of one page in half the time that it takes to copy a message two pages long, so can a half-length RNA molecule be copied in half the time by the replicase.

Suppose that the chain of viral RNA should be broken in half, by a random reaction with water that severs one of the phosphate-to-sugar bonds. Each half would now be a new individual, one which could be copied by the replicase in half the time. In practice, however, only one of the halves will be copied. We mentioned earlier that the replicase can distinguish $Q\beta$ RNA from the other RNA molecules usually present in a bacterial cell. It does this by attaching itself to certain sequences near one end of the viral RNA. These key sequences do not occur in bacterial RNA. Once bound to the sequences, the replicase is then ready to proceed with the copying process.

If a molecule of $Q\beta$ RNA were to split, only one fragment would receive the recognition sequences. The other would not be recognized; effectively it would be sterile. The recognized fragment would proliferate rapidly and come to dominate the mixture. Eventually one of its progeny would suffer a random chain break, producing an even more fertile descendant. This process would end only when the shortest chain was produced which can be derived from $Q\beta$ RNA and still contain the necessary recognition sequences.

In their early experiment, the Spiegelman group followed the evolution of $Q\beta$ RNA in the test tube for over 70 generations. At the end of this time, the mixture was dominated by a single RNA species, 550 nucleotides in length. Most of the now useless genetic information had been discarded, to obtain a more rapid replicator.

Another series of experiments was run with RNA that had already been shortened optimally. This RNA was allowed to replicate in the presence of a drug that slowed down the process. The drug bound to RNA at certain preferred sequences of nucleotides. As the moving replicase reached a site where the drug was bound, it was forced to push past it, just as we might have to shove aside a carton blocking our path in a supermarket aisle. Time was lost in the process.

A number of generations of the RNA were run in the presence of the drug, and the descendants were analyzed. Again, a single type of molecule was present. It differed from the starting RNA by three changes in nucleotide sequence. These changes had destroyed the favorite binding place of the drug. As a result, the replication rate had returned almost to the original level when no drug was present. The RNA had altered itself once again, in a manner that increased its reproduction rate.

These experiments, and others of a similar nature, demonstrated that a single molecule can adapt genetically to changes in its environment. For this reason the process has been termed evolution in the test tube.

The Skeptic wishes to intrude at this point. He reminds us that evolution involves a gain of new abilities, along with an increase in complexity. He asks whether the RNA has really evolved in this sense.

In the first experiment described above, notes the Skeptic, the RNA lost most of its original information. The drug experiment only demonstrated adaptation in the face of adverse environmental circumstances. The RNA did not acquire a new capacity, nor could it do so. With no access to protein-synthesizing machinery, it could not, for example, develop an improved replicase, or an enzyme that could destroy the drug.

For the advocates of nucleic-acid-has-primacy in the origin of life, this lack of control by RNA over proteins, including its own replicative enzyme, has been a vexation. One goal of research in this area has been the development of a system in which a nucleic acid can replicate without the aid of a protein. Some partial progress has been achieved.

Leslie Orgel and his colleagues at the Salk Institute in La Jolla,

California, have devised artificial energy-rich nucleotide sub-units. When these subunits were mixed with certain (not all) RNA molecules, they united and formed a new chain that matched the existing one according to the Watson-Crick rules. The single-stranded initial RNA was converted to a double helix, without the aid of a replicase.

The new double helix had some misconnections in its sugar-phosphate backbone, and the average length of the new chain was limited to about fifteen units. The special subunit that was used was found after a lengthy trial-and-error process, and was not one likely to exist on the early earth, according to Orgel. In addition, the process stopped when a double helix had been made. The RNA was not further replicated.

For these reasons, Orgel has been quite reserved in presenting his work, calling the reaction a model. Others have been less cautious and regard it as a clear indication that a naked gene could somehow, without protein, replicate itself on the early earth.

At the time of this writing, however, no demonstration has been made that a nucleic acid can manage without a protein. Certainly, the Qβ test tube system cannot function without the replicase. A different, and unexpected, result has emerged from this system, however. It is the RNA that is the unnecessary component.

Extensive studies have been performed on the Qβ test tube system within the last decade by Nobel laureate Manfred Eigen and his colleagues at the Max Planck Institute in Göttingen, Germany. In certain experiments they mixed the replicase, active nucleotides, and salts, but omitted the RNA. Nothing happened for a time, but after a lag, which varied from one experiment to another, an RNA appeared and then replicated and evolved.

The RNAs that emerged initially were a mixed population, with some as short as 60 units in length. During the subsequent evolutionary process, they gained in length and eventually a single species resulted, 150 to 250 units long. The actual sequence of this single winner varied from one experiment to another, however. In short, the replicase, when given no RNA to copy, constructed one on its own

This result was so unexpected that it was felt that a tiny amount of RNA must have been present, to start the process. Rigorous efforts were made in several laboratories to rule out this alternative. Even now, the argument is not quite settled. The possibility remains that a small amount of RNA was present at the start in each case. (By now I hope that the reader has grown accustomed to this state of uncertainty. Normal science works this way.) It seems likely that the result will hold up, however. If so, it is not one that gives comfort to the nucleic-acid-has-primacy group. We will return to it in a later chapter, when we give the protein viewpoint its due.

Manfred Eigen and his colleagues have constructed an elaborate theory of the origin of life, based in part upon extensive mathematical calculations, and also upon their results in the $Q\beta$ system. Their calculations have dealt with the interrelations among large molecules during the early development of life.

Their considerations start with a prebiotic soup. They do not use the ones that are produced by the best Miller-Urey experiment, but rather assume a much more nourishing biochemical broth of their own. Their recipe contains randomly made small proteins, lipids in sufficient abundance to form membrane fragments, and the natural active nucleotides or other energy-rich subunits suitable for the construction of nucleic acids. The critical event in this mixture is the assembly, by chance, of a molecule able to replicate itself. RNA rather than DNA is selected for this honor.

Despite the current celebrity status of DNA, there are a number of reasons which suggest that RNA preceded it in the origin of life. RNA certainly plays a more versatile role than DNA in life today. As we have seen, three different forms of RNA are essential in protein synthesis. RNA has a small but necessary part in bacterial DNA replication. In certain viruses, RNA functions as the genetic material itself. DNA is much more limited.

In the development of communities, handymen usually appear before specialists. This was probably true in the evolution of life as well—RNA came before DNA. Other indicators point the same way. In our current biochemistry, the building blocks of DNA are made from the corresponding RNA subunits. This may reflect the historical order of events. There is a reason for this as well. In chemical syntheses, the sugar component of DNA, deoxyribose, is more difficult to prepare, and more readily decomposed, than the RNA sugar, ribose. The discovery of deoxyribose in nucleic acids by chemists was delayed for just this reason. Deoxyribose probably was not present in any prebiotic soup, but was rather introduced into life processes when enzymes had evolved that could deal with it.

Let us return to the Eigen scenario. It presumes that life began on the day when one or more replicating RNA molecules were formed by accident in their enriched soup. This idea would certainly please a naked genie, but it deserves an identity of its own. In this account, the RNA molecule need not have been naked. It may have been given modest assistance in replication by proteins that had also been made at random and cohabited in their soup. Perhaps we can say that the RNA had a figleaf. Further, the RNA was not a gene. Like Qβ RNA in the test tube, it coded for no protein. It only replicated. We will call it the random replicator.

One or more RNA molecules were formed by chance. If only one was present, it soon diversified, due to inexact replication. In any event, after a period of competition and evolution, a winner emerged. As in the Qβ experiment, it was a molecule good at replication. It did not have a unique sequence, but consisted rather of a group of closely related individual molecules, termed a quasi-species.

Eigen and his collaborators have subjected that quasi-species to mathematical analysis. They feel that a length of 100 units of RNA could be attained, but beyond that, copying errors would destroy its identity.

The next phases have also been deduced from calculations, but are not specified in biochemical detail. RNA molecules, in some manner, learned to exert control over proteins and influence their composition and function. A primitive genetic code then developed. The different RNA molecules in the quasi-species took on different functions and cooperated for their mutual benefit. For example, a single RNA might serve to control each different amino acid (as transfer RNA molecules do today). Together, they could construct a protein.

A complicated and cooperative series of interactions, of checks and balances, developed between various nucleic acids and proteins. They have been named hypercycles and subjected to extensive mathematical analysis. The development of hypercycles took place in a continuous solution, uninterrupted by partitions of any sort. They filled up the soup. At this point, separate competing organisms did not exist. The hypercycles gained in complexity, and in their control of the environment, until a limit was reached.

For further progress to take place, it was necessary to reintroduce competition. The lipids that had been present in Eigen's benign broth were now utilized to construct compartments. Initially, the compartments contained similar contents. As random mutations exerted their effects, however, diversification took place. Differing hypercycles, each contained in its own membrane, competed with one another. Cells now emerged on the earth.

At this point, we can merge the account of the Eigen group with an earlier theory of Norman Horowitz. The earliest cells may have relied on the soup for their supply of construction units and suitable sources of energy. As cells proliferated, their numbers gradually exhausted the goodies made by prebiotic chemical synthesis.

Let us suppose that until now an important chemical had been made in the soup by the route $A \rightarrow B \rightarrow C \rightarrow D$. In this scheme, A was an abundant and inexhaustible material, such as a major component of the atmosphere. The early organisms required only the last product, D, for their important processes. Eventually, as the organisms multiplied, consumption of D exceeded its constant production and supplies became scarce. Competition for the limited available amounts of D stiffened, and survival became difficult.

Eventually, an organism acquired through mutation the ability to produce D from C internally, by enzyme-catalyzed pathways. This organism could grow using C instead of D. It proliferated, and dominated the environment. Ultimately, the competition either learned to make D from C or it simply died out, and the favorable mutant diversified further. Whatever the path, after a

time C was also depleted. A scramble then resulted until organisms learned to make C from B. This process was extended
backward, until the simplest resources were used in life processes. Ultimately, photosynthesis was developed. At that point,
some organisms could use solar energy directly, in addition to
the normal components of the air and soil. The soup was no
longer needed.

This combination of the earlier theory of Horowitz with the
current work of the Eigen group provides a coherent and fairly
continuous account of the origin and development of life from
soup to self-sustaining cell. There is a fair amount of disagreement about specific mechanisms and structures that may have
come into play after the first nucleic acid replicator was formed,
but this is set within a broader sense of confidence that the overall picture is finally coming into focus. Eigen and three co-
authors closed a recent *Scientific American* article with the words:
"The principles guiding the evolution of such an organization
have been formulated and experimentally verified. Now what remains to be discovered is just what the favorable molecular
structures were." In other words, there is still work to be done,
but we can see the light at the end of the tunnel.

One satisfying feature of the scheme is that a single, generally
accepted principle, Darwinian natural selection, is extended back
to the time of the first replicator. It is interrupted by certain periods of molecular cooperation in the early stages, but nonetheless,
it dominates the entire development of life.

The most important gap in these entire proceedings concerns
the steps prior to the appearance of the first replicator. Natural
selection does not apply, and we are left with only chance itself.
Spontaneous generation crawls out of the woodwork once again,
but in a more limited way. We are not asking for an entire cell,
but only for a single fragment, one molecule, the replicator. The
idea is actually not a recent one. A Harvard biochemist, L. T.
Troland (he was cited by Muller as a forerunner of his thinking),
wrote in 1914:

> Consequently we are forced to say that the production of the orig-
> inal life enzyme was a chance event. . . . The striking fact that the

enzyme theory of life's origin, as we have outlined it, necessitates the production of only a *single molecule* of the original catalyst, renders the objection of improbability almost absurd . . . and when one of these enzymes first appeared, bare of all body, in the aboriginal seas it followed as a consequence of its characteristic regulative nature that the phenomenon of life came too.

We need only substitute "nucleic acid" for "enzyme" and "replicative" for "regulative" in Troland's account to update it. Oparin's 1924 paper, as we have mentioned, also invoked chance to generate his first crucial structure: "a gel in a colloid solution."

In Wednesday's Tale in the Prologue, I paraphrased a modern popular account, by Robert Jastrow, of the chance creation of the replicator. Others have appeared recently. For example, Richard Dawkins wrote in 1976 in *The Selfish Gene:*

> Processes analogous to these must have given rise to the "primeval soup" that biologists and chemists believe constituted the seas some three to four thousand million years ago. The organic substances became locally concentrated, perhaps in drying scum round the shores or in tiny suspended droplets. Under the further influence of energy, such as ultraviolet light from the sun, they combined into larger molecules . . . in those days large organic molecules could drift unmolested through the thickening broth. At some point a particularly remarkable molecule was formed by accident. We will call it the *Replicator.* It may not necessarily have been the biggest or the most complex molecule around, but it had the extraordinary property of being able to create copies of itself.

Dawkins then continues along the lines put forward by George Wald. Such an event would be unlikely, but it only had to arise once in a billion years. "Actually a molecule which makes copies of itself is not as difficult to imagine as it seems. . . . The small building blocks were abundantly available in the soup surrounding the replicator."

We badly need the point of view of the Skeptic once again. Obviously, the chances for the spontaneous generation of a nucleic acid replicator are better than those for an entire bacterium. But the latter case was so hopeless that there is room for enormous improvement, and matters could still be hopeless. In the bacterial case, the equilibrium calculations of Harold Morowitz left us with a need to climb to the 100 billionth floor of our Tower of Numbers, yet we calculated that the maximum number of trials available on the early earth would take us only as high as the fifty-first floor.

Now how difficult would it be to put together the replicator at random? The minimal published estimates of its size propose a single strand of RNA of perhaps 20 nucleotides. To build this structure, about 600 atoms would have to be connected in a specific way, much less than the many millions needed for a bacterium. More trials would also be available for the purpose of building it, as less time and space would be needed for each trial. The replicase of Qβ can put together 200 nucleotides in a minute when copying an RNA chain. We will assume that spontaneous assembly would proceed at the same rate, in the most favorable case. Thus a replicator could be built in a tenth of a minute. Furthermore, the space occupied by a 20-unit replicator might be only one-millionth of the volume of a bacterium, so that many trials could be run for each used to make a bacterium. Considering these factors together, we can assume that a maximal number of 10^{59} tries at a replicator were available. We have reached the fifty-ninth floor of the Tower of Numbers, an improvement of eight levels.

But what are the odds? J. B. S. Haldane recognized that the chances of obtaining a self-replicating machine depended on the number of parts to it. If the number was small, there was no problem: "By mere shuffling you will get the letters ACEHIMN to spell 'machine' once in 5040 trials on an average." If you could shuffle at the rate of once per second, it would require only 84 minutes to run that many tries.

This analogy suggests that it should not be hard to put together a smallish replicator, so we must look more closely at it. We will stay with the metaphor of language, but set aside the letters on cards in favor of another much-used situation: the monkey at the typewriter. Let's call him Charlie the Chimp. Charlie is special. He never gets tired, and types out one line per second, completely at random. We can adjust the typewriter so that each line contains whatever number of letters we wish, and further, we can add or remove letters from the keyboard.

We will try a simple example. If we set each line to be seven letters long, and left only *a, c, e, h, i, m,* and *n* on the keyboard, how long would it take Charlie to type "machine"?

He will need longer than we did in the card-shuffling exercise, since he can use the same letter more than once. The odds are 1 in every 7^7 tries, or 1 in 823,543. At one try per second, it would require about nine and a half days for Charlie to make that many attempts.

Now let us give Charlie a normal keyboard with, say, 45 keys. The odds suddenly escalate to 1 in 45^7, or 1 in 370 billion tries. It would take Charlie (or his descendants) 11,845 years to run that many attempts. The word "machine" does not arise as readily as Haldane's first analogy would suggest.

Things get rapidly worse when we use longer messages. We will let Charlie try for a bit of *Hamlet.* The phrase "to be or not to be" has 18 characters, if we count the spaces as characters. The chances that our chimp will type this out are 1 in 45^{18}, or 1 in 6 x 10^{29}. At one try per second, it will take poor Charlie more than 10^{22} years to do that number of tries. Should the open model for the universe be correct, Charlie will still be typing away long after the stars have ceased to shine and all the planets have been dispersed into space through stellar near-collisions.

But now we have developed a real thirst for Shakespeare. We want our monkey to type out "to be or not to be: that is the question," which has 40 characters. The chances then become 45^{40}, or about 10^{66}, to 1. This is a number 10 million times greater than the number of trials maximally available for the random generation of a replicator on the early earth.

There we have it. If the chances of getting the replicator at random from a prebiotic soup are less than that of striking "to be or not to be: that is the question" by chance on a typewriter, we had best forget it. The replicator would have about 600 atoms. The chances of Charlie typing a 600-letter message (twice the size of this paragraph) correctly are 1 in 10^{992}.

Of course, atoms and letters, molecules and words, cannot be compared directly, and the number of possible organic compounds that can be formed from 600 atoms cannot be computed easily. We might suppose that only 10 types of atoms were prevalent on the early earth. With a 10-letter keyboard, the odds against a 600-letter message are "only" 10^{600} to 1. Furthermore, some fraction of these compounds could not be made, or would be unstable, for technical reasons. On the other hand, organic molecules are three-dimensional, exist in mirror-image forms, and have other complexities not present in linear English type. On the basis of simple chemical arguments, it can easily be shown that at least 10^{100} stable organic molecules, containing up to 300 atoms, can exist.

We could also use a very different approach to reach a similar conclusion. In an earlier chapter we considered the method of Harold Morowitz. He did not compute total possibilities in his approach, weighting all of them equally. Rather, he calculated what a group of atoms would prefer to do if they came to equilibrium. We cited his odds against getting a bacterium. For a small virus, we would need only to go to the 2 millionth floor of our tower. For a small enzyme, a trip to floor 8,000 would be necessary. He did not list data for a replicator in his table, but it would, by extrapolation, fall many hundreds or perhaps a thousand or two floors up.

In all of these methods, the odds against the random generation of a nucleic acid replicator still rest considerably above the chances, in the Tower of Numbers. They are still so unfavorable that the formation of the replicator by chance would seem miraculous (for a distance of even a dozen floors in our tower reflects odds of a trillion to 1, and a win in such circumstances would appear a miracle).

There is a further irony. Even should the miracle occur and the replicator find itself awash in the seas of the prebiotic earth, its fate would be unkind. It would perish without further issue. For in this random sea, it would encounter only hosts of unrelated chemicals, and not the subunits it needs to reproduce itself. A second miracle would be needed to surround it with exactly the ingredients it needs for further progress.

Having said all this, we have not yet joined the real issue with the nucleic-acid-has-primacy advocates. Most of them would probably agree with the analysis thus far. Certain popular accounts may imply that the first living molecule was formed from a completely random chemical mixture, but the scientists believe otherwise. If I may paraphrase their position, it would run something like this:

The first replicator was not formed from an equilibrium mixture. It was subjected to energy from various sources—lightning, solar radiation, and so on—and moved away from equilibrium. So the calculation of Morowitz is not relevant. As the energy influx continued, more and more complex compounds were made. However, all possible compounds were not formed. Certain ones were made preferentially, others hardly at all. The active subunits of the replicator, and other important biochemicals, were prominent among the compounds that were made. The replicator arose by chance, but with this mixture as the starting point.

I have expressed their position myself, for clarity. As this is a critical point, however, perhaps we should let the adherents speak for themselves as well. Manfred Eigen and his colleague Peter Schuster wrote in a 1978 article: "Here we simply start from the assumption that when self-organization began, all kinds of energy-rich material were ubiquitous, including in particular: amino acids in varying degrees of abundance, nucleotides involving the four bases A, U, G, C, polymers of both preceding classes . . . having more or less random sequences."

To reinforce this, let us quote the words of B. Kuppers, another German scientist of the Göttingen group: "Indeed, numerous experiments in the field of primordial organic chemistry demonstrate quite quickly that biological macromolecules (amino acids, energy-rich nucleoside phosphates) could form and polymerize spontaneously to proteins and nucleic acids." In this quote, read "large molecules" for "macromolecules," "active nucleotides" for "nucleoside phosphates," and "join together" for "polymerize."

If these claims were correct, then the origin of life would be a much easier business than it appears to be. Let us presume, for example, that a prebiotic soup had some forty biochemicals that were present in reasonable amounts and capable of uniting to form larger molecules. We will also assume that fresh supplies of biochemicals were furnished continually, so that trials could be run one after another, without any pause to acquire new material. Further, we will suppose that the natural active nucleotides (or equivalent subunits) constituted 10 percent of the entire mixture. If these presumptions were all true, then the chance of putting twenty nucleotide units together in a row would be "only" 1 in 10^{20}. These odds are still formidable, but they fall within a range that would permit us to win if we had a billion years at our disposal and a number of suitable locations in which experiments could be run.

Again, the Skeptic has demanded our attention. The analysis may be correct, he points out, but are the assumptions true? It has been difficult enough to establish that the early earth had a reducing atmosphere, and that even the simple amino acids were present. Why should we expect abundant, *ubiquitous* pools of nucleotides? No nucleotides, or even nucleosides, have been reported from the irradiation of simulated atmospheres in Miller-Urey experiments. None have been detected in meteorites, or seen in interstellar space. What experiments support the idea that the early earth was full of them?

To answer him we must explore a field called prebiotic chemistry, which occupies the attention of many experimental scientists concerned with the origin-of-life question. The prebiotic chemist sets up, runs, and analyzes the products of chemical reactions, an avocation that he shares with many other chemists who do not work on the origin of life.

The prebiotic chemist, however, operates under a self-imposed set of constraints. He is attempting to simulate reactions that may have occurred on the early earth in order to find a plausible series of steps that may have led to the origin of life. An ordinary chemist, when attempting to prepare some new substance, may select any reagents and conditions that serve his purpose. The prebiotic chemist limits himself to conditions that were present on earth before the start of life. Since these are unknown, the earth today is usually chosen as the standard, except for the atmosphere, which is presumed to be reducing in nature.

Synthetic chemists, unlike their prebiotic colleagues, can employ organic solvents such as ether, carbon tetrachloride, alcohol, and liquids derived from petroleum. As often as not, water is an enemy to be excluded as rigorously as possible. My own graduate training involved these techniques.

I can remember one aggravating ordeal from my first laboratory course in organic chemistry. The class had to carry out a procedure named for the Nobel Prize–winning French chemist Victor Grignard. It required the combination of a metal, ether, and an organic compound in a protected flask. But any trace of moisture would ruin the reaction. No breath, no touch of saliva, no common whiff of laboratory air was permitted within the apparatus lest the procedure be ruined. Flames, airtight seals, protective tubes of chemicals that devoured water greedily, all were employed in a determined effort to maintain the anhydrous virginity of the contents of the flask. Only when these rigorous and unnatural demands had been met was the Grignard reaction willing to give its benediction and proceed. It signaled its assent by releasing a stream of bubbles at the shiny surface of the metal. When I ran the procedure it was only then, when the mixture grew agitated, that my own nervous system could calm down.

The prebiotic chemist is excused from this ordeal, though he would probably be quite willing to have it. Whatever disagreements may exist about conditions on the early earth, there is consensus on the presence of abundant water. No reasonable prebiotic simulation can exclude it fully, which is unfortunate, for practical reasons. For we have seen that the subunits of our large molecules are put together in a process that involves the formation of water. Whenever two amino acids unite, a water molecule is released. Two molecules of water must be set free in assembling a nucleotide from its components, and additional water is released in combining nucleotides to form nucleic acids.

Unfortunately, the formation of water in an environment that is full of it is the chemical equivalent of bringing sand to the Sahara. It is unfavorable, and requires the expenditure of energy. Such processes do not readily take place on their own. In fact, the reverse reactions are the ones that occur spontaneously. Water happily attacks large biological molecules. It pries nucleotides apart from each other, breaks sugar-to-phosphate bonds, and severs bases from sugars. These reactions are taking place in our cells at this very moment. Fortunately, after billions of years of evolution, our bodies are well equipped to deal with these events. We have developed elaborate mechanisms to repair the damage to our molecules caused by continual attack by water.

On the early earth, such defenses did not exist. Water continually opposed the assembly of large biomolecules and attacked those that had successfully formed. Yet it is the task of the prebiotic chemist to demonstrate that such molecules could be formed. Much as he would like to, he cannot employ Grignard-type conditions. He must settle for less evasive maneuvers.

Restraint must also be practiced in the selection of reaction temperatures. The earth today serves as the model. The conditions used in prebiotic simulations may vary from Saharan heat to Siberian cold. This range, though wide, is still limited when compared to the one available to the ordinary chemist, who does not hesitate to employ molten salt and liquid air.

Finally, we must discuss the use of acid and alkali, another pair of opposites, a yin and yang of chemistry. Although they oppose and destroy each another, they share a common lack of sympathy for our life substances and for materials derived from them, such as our clothing.

In my early laboratory days, the telltale holes in my trousers or (when I was wiser) in my laboratory apron would testify to my lack of respect for acid or alkali. Fortunately, when I got one of them on my skin, my nervous system sent an urgent message that I had best wash the offending substance off as quickly as possible, and I heeded the signal. Our tissues do not like acid or alkali. They prefer the balanced state between the two extremes, aptly named neutrality. Chemists use a numerical scale, called pH, to keep track of these quantities. In this system, the number 7 represents neutrality, lesser numbers are the acidic range, and everything above that is alkaline.

The pH scale is constructed on a logarithmic base, as is COSMEL and the Tower of Numbers. Thus, a solution of pH 6 has 10 times the acid content of one of pH 7, and a solution of pH 5 is 10 times more acidic yet than one of 6. A solution of pH 10, three steps removed from 7, is 1,000 times more alkaline. In neutral solution, both acidic and alkaline properties are present to a slight extent, but they exactly balance each other.

Pure water left to itself will be neutral. A sample of water becomes acid or alkaline when substances that possess acidic or alkaline qualities are added to it. Thus vinegar contains acetic acid, and our stomachs contain a weak solution of hydrochloric acid. The common household cleanser marked "ammonia" is an alkaline solution of the gas ammonia in water.

The biochemical processes typical of life on earth prefer neutral conditions. Our blood maintains a pH of about 7.4, called physiological pH. Most reactions of enzymes, and other processes that take place within our cells, occur optimally near neutrality. Mild departures may be tolerated, but presumably they cause an imbalance in these processes.

Strongly acidic or alkaline conditions are quite harmful. The weaker bonds that help our important molecules keep their shape are disrupted, and the rate at which water inflicts more permanent damage on these molecules is accelerated. Such reactions, for example, made the holes in my trousers.

Nonetheless, microorganisms exist that tolerate these circumstances. Bacterial strains related to the methanogens may survive in alkaline lakes of pH 11 or hot acidic springs of pH 1. The cells lining our stomach can withstand the pH 2 solution within it. These cells do not survive and thrive by adopting a different chemistry, but rather by protecting themselves against the environment. They pump out the offending acid or alkali, as a refrigerator might pump out heat, and thereby preserve conditions closer to neutrality inside.

The pH of the seas of the early earth is not known, but is usually assumed to be not far from neutrality. The prudent prebiotic chemist will therefore limit his conditions to those near a pH of 7. The acid springs and alkaline lakes provide an excuse for the use of more extreme pH values. However, their extent is quite limited, when compared to a vast prebiotic ocean, so there is a great loss in probability, in the number of possible trials, when they are invoked. We can also reason that a type of life that functions best today under neutral conditions is more likely to have originated and developed under similar conditions.

With these various constraints in place, it then falls to the imagination of prebiotic chemists to devise a series of plausible reactions that demonstrate how an initial simple chemical mixture will afford important biomolecules. In the particular case of the random replicator theory, the goal has been to generate an environment rich in the subunits of nucleic acids, one suitable for the production of the replicator itself by chance.

Have they succeeded? Some very competent and skilled chemists have worked in this area. They have used ingenuity in devising the reactions, and care in analyzing them. With few exceptions, their results have been accepted as correct. The interpretation of them is another matter. Once again, as in the case of the Miller-Urey experiments, we must examine the details closely.

The prebiotic synthesis of the replicator starts with a reducing atmosphere of the type used by Miller. The amino acids are not of interest here, but simpler intermediates formed initially in this atmosphere: hydrogen cyanide and formaldehyde. These two substances have the same relation to primordial reaction recipes that olive oil and tomato sauce have to southern Italian cuisine.

Both intermediates have few atoms. Hydrogen cyanide has one atom each of hydrogen, carbon, and nitrogen, and is the simplest compound that can be prepared from these elements. Formaldehyde is the smallest molecule preparable from carbon, oxygen, and hydrogen, with one atom each of the first two elements and two of hydrogen.

Ironically, though these two molecules are often invoked in the origin of life, they are used as substances of death today. Formaldehyde is used as a preservative for the storage of specimens in biological laboratories, and hydrogen cyanide is the lethal agent of execution in states that use the gas chamber.

These two chemicals function well in both roles, life and death, because of their substantial reactivity. They happily combine with water, one another, and many other chemicals. When deprived of better alternatives, each combines with itself. These chemicals form transiently when reducing atmospheres are exposed to an appropriate energy source, then react in various ways to form Miller-Urey products.

Prebiotic chemists start with this observation and then employ a procedure that is characteristic of this field. They assume that a substance, once demonstrated in any amount as a product of a prebiotic reaction, can then be employed in pure form and enhanced quantity as the starting material in a very different prebiotic transformation. This process is repeated until an entire series of reactions has been put together, to connect the reducing atmosphere with a replicator.

To present the reasoning behind this procedure, I have devised an imaginary prebiotic spokesman and named him Dr. Midas, after the legendary king whose touch turned ordinary substances into gold. Dr. Midas, in the same tradition, can convert mundane chemicals into genes with a wave of his hand and a well-chosen phrase.

We will let him perform, and describe the route to the replicator: "Cyanide and formaldehyde were formed in the primitive atmosphere," he notes, "so we will start with them." He points out that formaldehyde, when exposed alone to appropriate conditions, forms a mixture which contains some of the sugar ribose. Hydrogen cyanide, under very different conditions, is converted in part to adenine, one of the important bases of the nucleic acids. The other bases can also be prepared, but by longer and more indirect routes. "So you see," says Dr. Midas, "we can make the bases and ribose. The next problem is the nucleosides."

Adenine and ribose can be heated together with appropriate catalysts to yield a mixture containing adenosine, a nucleoside component of RNA. The conditions again differed from those used in the earlier steps. "So much for the nucleoside problem," says Midas. "Now let's do the nucleotides."

When adenosine is heated with phosphate and different catalysts than those used before, a natural nucleotide is among the products. Midas comments: "We have shown that nucleotides could have formed on the prebiotic earth. We must now combine them to form a nucleic acid."

Still other procedures, starting with nucleotides, have in fact shown that a few units can be connected together. "That clinches it," Dr. Midas concludes. "We know that single-stranded nucleic acids can be converted to a double helix, when appropriate sub-units are supplied. Obviously, there was no real problem in as-sembling a double helical nucleic acid on the primordial earth."

The Midas point of view has been put forward many times, and made into a sub-paradigm of the current origin-of-life theory. For an example, we can turn once more to A. L. Lehninger's 1975 biochemistry text. In the quote, the terms "pyrimidines" and "purines" refer to base subclasses, and "deoxyadenosine" is a nucleoside component of DNA:

> ... the organic building blocks of nucleotides, namely pyrimidines, purines, ribose and 2-deoxyribose have been shown to arise under simulated prebiotic-earth experiments. Moreover, nucleosides like adenosine or deoxyadenosine have also been detected in simulated primitive-earth experiments. When nucleosides and polyphosphates are heated or irradiated with ultraviolet light, mixtures of nucleotides are formed. . . . The next step in the chemical evolution is the formation of internucleotide linkages between successive nucleotides. This has also been observed in simulated primitive-earth experiments . . .

The Skeptic has been silent during this presentation, but he has grown increasingly agitated. Now he speaks.

"These experiments show only that a chemist could prepare a nucleic acid in the laboratory today, using a variety of conditions that he chooses to call prebiotic. Even this preparation is not carried out in a continuous manner. Formaldehyde is not collected from a Miller-Urey experiment, purified, and used to make ribose (though undoubtedly this could be done, if modern equipment was employed). Instead, formaldehyde is simply detected as an intermediate in the atmosphere, then the pure chemical is bought from a supply house and used in the next reaction. This type of practice is followed at every step down the line. Unfortunately, on the primitive earth, there was neither modern equipment nor supply houses, and certainly no chemists."

Dr. Midas responds: "Of course we have taken some short-cuts, to save time. We are only human, and do not live forever. We wished to demonstrate, in a few weeks, the steps that took a billion years on the early earth."

The Skeptic now asks Dr. Midas whether the availability of a billion years is enough to justify this procedure, citing our earlier example of the monkey at the typewriter. He takes Midas over to the corner where Charlie the Chimp is still happily banging away on the machine, and asks: "How long do you think it will take for the chimp to type 'to be or not to be: that is the question'?"

Midas picks up a line of random script and examines it. "Not very long at all. Look, there's a *t*, and further down the page there's an *o*, and so on. All of the necessary steps could be done."

"But can the letters by typed in the right order?" asks the Skeptic.

"No problem. I only need the proper materials."

Midas departs and returns with a bunch of bananas and a fresh pad of typing paper. He removes Charlie, and types for a few minutes, changing sheets frequently. He then places the monkey in the typist's chair once again. He has set the typewriter so that it moves to a new line each time a letter is typed.

The monkey starts to type, with Midas watching over his shoulder. "Aha!" Midas yells after a few seconds, stopping Charlie. He gives him a banana, pulls the sheet from the typewriter, and shows it to us. About two dozen letters have been typed, each at the start of a line. The last of them is a *t*.

"We've shown that the monkey *could* type a *t* to start a line," Midas claims triumphantly. "Now we'll try for an *o*."

He pulls a sheet of paper out of his pad. He has typed a *t* at the start of every line. He puts this sheet into the typewriter, sets the margin so that the next letter struck on each line will fall to the right of the *t*, and turns the monkey loose again.

After about half a minute, he shouts and interrupts the monkey. Once more he brings the sheet to us. Each line now contains a two-letter unit starting with *t*. The first thirty are meaningless, *tx*, *t!*, *te*, *tt*, and so on, but the last one is *to*.

"There!" says Midas. "The monkey has typed the word 'to.' Now we must try for the space."

Thoughtfully, he has prepared in advance a sheet with the word "to" typed at the start of each line. He returns Charlie to the typewriter.

An hour and a half later, after a number of such operations,

Midas is ready to insert the last sheet into the typewriter. This one contains the message "to be or not to be: that is the questio" at the beginning of each line. Charlie dutifully types away, adding a different letter at random to each line, until he produces an *n*, upon which Dr. Midas rewards him again, and stops the process.

"There is the line you wanted," he concludes. "I've shown you that the monkey could do it. I've speeded the process up a bit, but that's because I've other errands to run today. But it is possible. The monkey, left to himself, would just need a while longer. Give him enough time, and you'll surely get the message."

Midas departs, bowing gracefully.

"And that shows you what *I* mean," the Skeptic says. "The monkey didn't type the line. Midas did it. He cued the monkey every time, when the correct letter was typed, and then gave him a fresh start each time, with all the previously typed, correct letters collected together.

"Prebiotic chemists do the same thing. They run a lot of reactions until they get the compound they want. Once they have done this, no matter how many trials they needed or how low the yield of the desired product, they feel free to go to the next step. In doing so, they start with a fresh, pure supply of the compound they've made. They claim that they must cut a few corners to save time.

"But look at the size of the corner that Dr. Midas cut with Charlie. The chimp needed about 45 seconds to strike each letter at random. For the 40-letter message, the total monkey typing time was 45 times 40 seconds, or 30 minutes. Left alone, he would have faced odds of $(45)^{40}$ to 1. As we saw a while ago, he probably would have needed 10^{59} years or so to get the message right (though if he were very, *very* lucky, he could of course get it on the first try). Not a bad trick to substitute 45 times 40 for $(45)^{40}$."

The Skeptic has concluded, but I will add some historical notes to his argument. Experimenters in many areas of science have provided unconscious cues to their human or animal subjects, or even built them into the design of their apparatus. One famous case, cited in psychology textbooks and also told well by

Carl Sagan in *Broca's Brain*, concerns the mathematically adept horse, Clever Hans.

Hans lived in Germany at the turn of the century and was famous for his ability to do arithmetic. For example, his owner might instruct him to add 14 and the square root of 4 then subtract 5 from the sum. Hans would start to tap his hoof slowly, and stop after 11 taps, signaling the correct answer. Hans's owner would then reward him with a lump of sugar and a hug. If the owner had persisted, he could obviously have induced Hans to perform the entire computation of his income tax.

This ability was amazing, but unfortunately, Hans lost it when the owner was unaware of the answer or kept out of Hans's sight. The owner, quite unconsciously, had signaled to the horse when to stop tapping by changes in his body tension. The horse had learned that if he stopped at this cue, he would get his sugar.

We must return to the question of the replicator. Could it have arisen at random in one billion years on the earth? Prebiotic chemists are correct in their claim that the work of a billion years cannot necessarily be duplicated in an afternoon. On the other hand, this disclaimer cannot be used to give validity to reaction sequences of monumental implausibility.

The advocates of the random replicator have not presumed that the synthesis of nucleotides was rare but rather that these substances were plentiful in prebiotic times. The step that required chance was the combination of nucleotides into a nucleic acid.

If this were so, it should be easy to demonstrate the abundant synthesis of nucleotides from the components of the primitive atmosphere and soil. Ideally, we would want to mix the right ingredients together, seal the flask, set it aside for a few hours or days, and then harvest the rich nucleotide crop.

This has not been done. The various stages have been run separately, in poor yield, under very different conditions. As they have not been combined in fact, we shall now unite them in fantasy. I have linked together some of the most cited prebiotic syntheses in the nucleic-acid field into a continuous narrative. The suggestions of the experimenters have been used as much as possible, but where they are lacking, I have put in details of my own. The end result is presented in the form of a fable, an amended version of Wednesday's tale from the Prologue. For those who wish to consult the technical papers used to compile this narrative, I have provided a list in the references at the back of the book.

Wednesday's Tale (Revised)

Once, a long time ago when the earth was quite young, a group of high mountains rose above the ocean, forming a large island. It was volcanic, somewhat like a Hawaiian island of today, for continents as we know them had not yet formed. Because of the height and extent of these mountains, and because of the prevailing wind and weather patterns, a variety of climate zones existed on the island.

Thunderstorms were frequent on the rainy side, where it was always cloudy. In the high altitudes, near the mountaintops, the rain froze, and the precipitation came down as snow or hail. The atmosphere was reducing, and these conditions favored the formation of hydrogen cyanide in the discharges. The rain and snow were rich in this chemical.

Large glaciers descended from the highest peaks. At their base, in the summer season, lay a number of partly frozen alkaline lakes. Hydrogen cyanide collected in them, and reacted with itself extensively, until the time came when the lakes froze solid in the winter. When warmer weather resumed, the lakes thawed in part and the reaction started again. In one very important year, however, spring did not return. The climate in the highlands had taken a turn for the worse. More snow fell at the mountaintops and the glaciers advanced, pushing the frozen lakes down the mountain. The flow path of one glacier led it away from the wetter side of the island toward a central plateau, which was geothermally active. In this more temperate climate the glacier tip melted, and the hydrogen cyanide reaction mixture flowed into a boiling acidic spring.

Such boiling springs exist today in areas like Yellowstone Park and Iceland. Bacteria, which belong to the same broad class as the methanogens, are able to grow there. In the early days that we are considering, of course, no life existed, but over the course of an hour the boiling acid converted a small amount (about 0.1 percent) of the solids that the glacier had brought into adenine. The acid would eventually also have destroyed the adenine, but before that could happen the spring waters flowed into a broader stream. In doing so, they passed over some alkaline soils which neutralized them.

It seldom rained in this broad plateau area, and when it did, it fell in the form of sunshowers, rather than thunderstorms. The rays of the sun caused formaldehyde, rather than hydrogen cyanide, to be formed. The formaldehyde rain flowed in tiny streams into a geologically different, but also geothermally active, part of the central plateau, which contained boiling neutral pools, thick with suspended minerals.

As each formaldehyde stream flowed into a boiling mineral pool it was converted into a complicated mixture by a process called the formose reaction. The sugar ribose formed a small part of this product. Moving waters carried the mixture down the length of the pool over the next several hours, allowing enough time for the change to be completed. At this point the product flowed out of the hot pool and was swept downstream by a rapid icy brook. This escape was fortunate, as the ribose would have decomposed if it had remained too long in the pool.

The adenine and ribose streams merged in the central plateau, but they could not yet form adenosine. They needed a hot environment and the presence of sea salt for that purpose. Happily, a precipitous waterfall took them almost to sea level, on the hot, dry side of the island. Time was of the essence, as the sugar was not stable and was being lost.

At the base of the waterfall, the stream widened to form a broad delta. The waters flowed over a variety of different types of rock and mineral formations. At some point they entered a tidal pool which had been cut off from the sea at low tide. Minerals lining the pool had a special affinity for both adenine and ribose, and retained them, while most of the other substances were swept away as the tide filled and drained the pool.

It was a very hot day. The sun evaporated the remaining water in the pool and heated the adenine and ribose in the presence of salt, converting them in part to the nucleoside adenosine. As this was happening, a violent storm occurred far out at sea, creating large waves.

The tides returned to the tidal pool in a rush, sweeping out its contents and transporting them farther inland. They were deposited in a nearby pond, which we name Darwin Pond. This was to be the chosen site for the origin of life.

No sooner had the adenosine reached Darwin Pond when successive waves, each flowing from a different direction, brought in supplies of the other nucleosides needed to make RNA. Had these chemicals only been human, they would have embraced at the joy of their first meeting, and in anticipation of the glorious future that lay ahead of them. They would then have taken turns, each describing the marvelous and different series of events that had led to its own creation. We must not inject our own feelings into the story, though. Let nature continue the synthesis.

Phosphate was needed for the conversion of nucleosides to nucleotides. Several geologists have contended that phosphate was not readily available on the early earth, and only increased in concentration in the waters gradually, as appropriate rocks weathered. Darwin Pond, however, was one of the few choice locations blessed with the right kind of mineral; it already had abundant phosphate. Thus, when the continuing heat wave evaporated the pond almost to dryness, the nucleosides were converted to nucleotides. This process was aided by an additional catalyst which was found in the minerals lining the pond.

The nucleotides now needed to combine, to form the replicator. This process was helped greatly by the presence of certain chemicals called amines which were brought in by another temporary flood. The amines would have been unwelcome earlier in our account, as they would have interfered with several earlier steps.

The climate now stabilized. Days were as hot as before, enough to dry up the pond. Each night, however, winds brought in enough moisture to form a thin liquid film at its bottom. These alternative periods of heat and moisture afforded the nucleotides a chance to come together in various ways and then to break apart again. One evening, by chance, the replicator was formed. It took charge immediately, assembling other nucleotides into copies of itself, more rapidly than they could come apart. Life had been created and evolution could begin.

Before ending this tale, we must comment on the name of the pond. Charles Darwin himself did not extend his theories to the question of the origin of life, and publicly identified himself with a belief in Creation. In 1863, in a private letter to the botanist Joseph Hooker, he wrote that "it is mere rubbish, thinking at present of the origin of life; one might as well think of the origin of matter." Yet he himself could not resist the temptation of playing with such rubbish, for he wrote in 1871, again to Hooker:

> It is often said that all the conditions for the first production of a living organism are now present, which could ever have been present. But if (and oh! what a big if!) we could conceive in some warm little pond, with all sorts of ammonia and phosphoric salts, lights, heat, electricity, etc. present, that a protein compound was chemically formed ready to undergo still more complex changes, at the present day such matter would be instantly devoured or absorbed, which would not have been the case before living creatures were formed.

This quote is often reproduced in texts and articles on the origin of life. Many workers would prefer to replace the word "protein" with "nucleic acid," as we have seen. Otherwise, it is remarkably current today, which is a tribute either to his foresight or to our lack of progress.

The Skeptic, who had looked ill earlier in the chapter, has recovered during the tale and is in fact rolling on the floor with laughter. He stops to ask how much of the tale I have composed and how much actually has been published in the scientific literature.

I respond that the very different reaction conditions have been published, as well as suggestions for appropriate prebiotic locations, such as the frozen pond, the boiling mineral pool, the tidal pool, and the dry, desertlike environment. I had to devise most of the transport to move the chemicals between locations. The glaciers and the separate rains of hydrogen cyanide and formaldehyde have also been published, however.

"It is very imaginative," he says, "but frankly, for a fairy tale, I prefer Father Raven."

Different accounts leading to the origin of the replicator could be constructed, using other experiments published in the literature. Some would be less spectacular than the above one, but all would share the same general defects. Many steps would be required which need different conditions, and therefore different geological locations. The chemicals needed for one step may be ruinous to others. The yields are poor, with many undesired products constituting the bulk of the mixture. It would be necessary to invoke some imagined processes to concentrate the important substances and eliminate the contaminants. The total sequence would challenge our credibility, regardless of the time allotted for the process.

Once more, as in the case of the Miller-Urey experiment, we have a large gap between the uncontested results of a series of studies and the myths derived from them. Again, we must inquire into the attitudes behind the belief systems that are involved. We can start with a statement taken from Zubay's 1983 biochemistry text: "The first life forms probably contained nucleic acids for the storage of genetic information. . . . Consequently there must have been a pathway by which the nucleotide building blocks of RNA were synthesized."

Although it is partly redeemed by the author's use of "probably," the use of "must" in the quote puts it more into the realm of mythology than science. It ignores the possibility summarized in this alternative statement: Determined efforts have failed to turn up pathways suitable for the production of abundant supplies of nucleotides on the early earth. Consequently, the first life forms probably stored their genetic information in some chemical system simpler than nucleic acids.

Yet the belief persists that such pathways must exist. Perhaps the foremost believer is Professor Cyril Ponnamperuma, who heads the Laboratory of Chemical Evolution at the University of Maryland. According to Ponnamperuma: "Nobody doubts now that the components of nucleic acids can be made by a path that can be called natural." Perhaps a bit more organic chemistry should be done to smooth out some difficulties, but this will surely take place. The pathways did not operate at random: "There are inherent properties in the atoms and molecules which seem to direct the synthesis in the direction most favorable" for the molecules of life.

He made these remarks when I interviewed him in his laboratory, actually a series of laboratories brightly decorated with posters from the space program, a display of a meteorite fragment, photographs of the Miller-Urey type of apparatus, as well as the apparatus itself, and a can marked "primordial soup." Before assuming his present position, he had worked at the NASA Ames Laboratory in California, where he had played an important role in the analysis of organic compounds in meteorites. Ponnamperuma is perhaps the best known of living scientists who devote their full time to the study of the origin of life. He was the first recipient of the recently created Oparin Medal of the International Society for the Study of the Origin of Life, and is the president of the society at the time of this writing.

Ponnamperuma comes from Sri Lanka (Ceylon). He had studied religion at the University of Madras and then moved to London to learn chemistry, according to interviewer Harold T. P. Hayes. "In the late 1940's, Ponnamperuma realized that his two quite disparate interests were intersecting" in the field of the origin of life, wrote Hayes. As we have noted, his approach to the area is pervaded by an optimism and a sense of cosmic purpose that seems to come from some inner faith.

This purpose begins in outer space. He stated his feelings eloquently in a recent comment: "You look at the interstellar molecules and you see cyanide and formaldehyde. These two can provide the pathway for everything else. There is a simplicity in the whole scheme—so much so that you practically feel that the whole universe if trying to make life." Because of the operation of these favorable factors leading to our own chemistry, "we are the brothers and sisters of the stars." In the Hayes interview, Ponnamperuma commented: "I wouldn't be surprised if you landed on some planet like earth and somebody about five feet two with two eyes came up to you and said hello." His most succinct summary was made in a recent public lecture: "God himself must be an organic chemist."

This viewpoint is one we met in our discussion of the Miller-Urey experiment and called predestinist. As we noted then, we can't exclude the possibility that the laws of the universe have been rigged in our favor. No greater compliment could be paid to us. At present, though, this attitude must be based on faith, as the evidence does not support it. The interstellar molecules, for example, can provide the pathway for everything else, as Ponnamperuma suggests, but I would emphasize the *everything*. All of Beilstein could be constructed, ultimately, from the molecules present there. When we consider the theories of Sir Fred Hoyle, we shall learn how far we can go, if we let our imaginations be provoked by these substances.

A much more somber appraisal of prospects in the naked-gene hypothesis can be had from Leslie Orgel. We have encountered him before, in connection with his model system for studying the replication of RNA without a catalyst. He is co-author with Francis Crick of a paper on directed panspermia, a topic that we shall consider shortly. He has made theoretical contributions to many areas of biological science, on topics that range from aging and mutation theory to selfish DNA. More relevant for our purposes is the fact that much of the best work on prebiotic synthesis of nucleic acids has come from his laboratory at the Salk Institute in California. In summarizing the state of this research, Orgel has been quite candid: "The formation of sugars in plausible conditions and their incorporation into nucleosides have not been achieved. Until this problem is solved or bypassed, it remains a weakness in theories of abiotic nucleic acid synthesis. The origin of nucleosides and nucleotides remains, in our opinion, one of the major problems in prebiotic synthesis."

When I met up with him at a Detroit conference in 1983, Orgel was ready to admit that the difficulties in the prebiotic synthesis of nucleic acids were overwhelming. But he was quite quick to add, "There are equally overwhelming difficulties in the way of all theories." He knows the work in the field in great depth, and on most points he could serve as an admirable substitute for the Skeptic who has accompanied us in this book. But then suddenly he can take a different turn: "I suspect there is a trick, but I don't know what the trick is." He likes to point out that most of the rubies on earth occur in one hill in Burma, the product of an unexpected series of transformations in mineral chemistry. We cannot know every kind of substance that may have been abundant on the primitive earth. Perhaps one of them, some magic mineral, had exactly the right properties to cause the necessary reactions to occur to create a nucleic acid.

As Leslie Orgel suggests, there may be some easy solution that has been overlooked. An immense number of combinations of minerals with other chemicals remains to be tested, and perhaps one of them will do the trick. Until that combination turns up, however, the idea of a naked nucleic acid gene must be considered either a speculation or a matter of faith, according to the attitude of the person who proposes it. In the interim, it will be worth while to consider other solutions, and we shall do so.

For a fitting summary of this topic, I would like to cite Graham Cairns-Smith, whose own theory shall get our attention a bit further on.

> There have indeed been many interesting and detailed experiments in this area. But the importance of this work lies, to my mind, not in demonstrating how nuleotides could have formed on the primitive Earth, but in precisely the opposite: these experiments allow us to see, in much greater detail than would otherwise have been possible, just why prevital nucleic acids are highly implausible.

If that conclusion is correct, then life employed some other genetic system before the advent of nucleic acids. No evidence exists concerning its identity, but there has been no shortage of speculations to provoke our thoughts. We will move on to consider some of the more prominent ones.

eight

Bubbles, Ripples, and Mud

Evolution requires a molecular system capable of both storing information and providing occasional variants as sources of possible improvement. We understand how modern nucleic acids function in this role. Unfortunately, their chemical complexity makes it unlikely that they were formed by spontaneous processes, and present at the start of life. What other system, then, filled this function in earliest times? This is a critical question for the origin of life.

There has been no shortage of suggestions, of course. Of all the competing theories, none has carried the day. Each theory has lacked the decisive demonstration that would convince its opponents, or at least, the uncommitted observers. The theories, though very different in detail, have another feature in common. Each starts with its own distinct view of the early earth. As I read these descriptions successively during the course of my research, they merged in my mind, affording a composite picture of our planet in its first years. It was an eerie, barren place, a wasteland of bubbles, ripples, and mud. Each of these items represents a distinct idea for the origin of life, and we will consider them separately.

Bubbles

Picture first a sea or lagoon filled with tiny bubblelike structures of bacterial size. They are not bacteria, however, since they are made only of proteins, or to be more exact, proteinoids, a related substance made by heating amino acids together. Despite this limitation in their composition, these structures, called microspheres, show numerous lifelike properties. They catalyze chemical reactions and possess surfaces that resemble membranes. Under certain circumstances they can produce electrical responses that resemble those of modern nerve cells. Above all, they can proliferate, and have the capacity to evolve by natural selection. As they do so, they produce for the first time those chemicals crucial to life today, proteins and nucleic acids. This is the vision of Sidney Fox, who heads the Institute for Molecular and Cellular Genetics at the University of Miami.

For the past quarter of a century, Professor Fox has been the most noted advocate of the proteins-first position on the origin of life. He has attacked the naked gene concept vigorously, claiming that "DNA arose from the living system; it was not the result of a separate act of special creation . . . the hereditary molecule DNA had to arise in cellular systems in which ordered proteins already existed."

He is now past seventy but still active in research and in organizing symposia that advance this point of view. In 1983, for example, he presided at a session of the Detroit meeting of the American Association for the Advancement of Science that featured ecologist and sometimes presidential candidate Barry Commoner as a speaker. This meeting drew headlines such as "DNA Role Downplayed" in the *Detroit Free Press* and "Research Builds Case for Late Evolution of DNA" in *Chemical and Engineering News*. This view, of course, is one that I also found attractive in a previous chapter.

Sidney Fox has not merely served as a rallying point for the proteins-first group, but has advocated the particular system of proteinoid microspheres, first demonstrated in his laboratory in the late 1950s, as *the* solution to the origin-of-life problem. Needless to say, this position has made him a center of controversy. His system has received favorable attention in the media and in a number of texts, most notably A. L. Lehninger's widely used *Biochemistry*, which termed it remarkable. On the other hand, it has attracted a number of vehement critics, ranging from chemist Stanley Miller and astronomers Harold Urey and Carl Sagan to Creationist Duane Gish. On perhaps no other point in origin-of-life theory could we find such harmony between evolutionists and Creationists as in opposing the relevance of the experiments of Sidney Fox.

When I interviewed the controversial Professor Fox during a recent origin-of-life meeting, I found him to be courteous, open, and quite generous with his time. He was willing to discuss the history of his own efforts, and of the origin-of-life field in general. When I asked him about the criticism directed at him, Fox recalled an incident when a close friend was preparing a personal history: "He had felt for many years that I was much too touchy about criticisms that I'd received." But then "he went back over some of the papers and he found that he had changed his mind and decided I wasn't touchy enough." Fox noted with regret, "I would like to think that science is just one huge pleasant fellowship . . . but it isn't like that, people get their emotions invested."

Why would he arouse these responses? Perhaps because he felt that he had largely solved the origin-of-life problem. Fox referred to another scientist who had published an extensive theory which outlined the important questions yet to be considered: "How is he going to feel after he finds out that we've answered these questions?"

Fox did feel very confident about the validity of his answers. He commented, "I think that evolution has followed a very narrow pathway which is very nonrandom. So one is either on the evolutionary track or he isn't." He cited his own laboratory, and perhaps half a dozen others, that were on the right track. "The others are all in a context of DNA first or of randomness, which are related ideas. They have dominated the picture, but I don't think that they have gotten us anywhere."

The Skeptic must intrude at this point in our narrative. He points out that whatever the interpersonal feelings that may be involved, the value of this system must ultimately be determined by the experiments themselves. So we must turn to the details.

We have seen that many experiments, starting with that of Stanley Miller and Harold Urey in 1953, have demonstrated the ready formation of certain amino acids under possible primordial conditions. Mixtures of amino acids do not have lifelike properties. When united into long protein chains in very specific ways, however, the amino acids can form enzymes. Enzymes are vital components of living systems, as they greatly increase the speed of chemical reactions important to life processes.

Amino acids do not readily unite to form peptides (short protein chains) and proteins when water is present. The details of the energy budget in fact dictate that the opposite should take place. In the presence of water, peptides and proteins slowly break down into amino acids. The situation suggests its own remedy. To unite amino acids, heat them together in a dry state, so that the water released by their union is expelled.

This remedy, when tried, had been found wanting, however. "Biochemists knew that when a mixture of amino acids in the ratio found in proteins was heated, the result was pyrolysis to a dark brown tar with a disagreeable odor," commented chemist William Day. At this point Sidney Fox made a contribution. Fox set aside the usual recipes and added extra amounts of any of three special amino acids. These mixtures, when heated in the dry state well above the boiling point of water, gave clean preparations, in which the amino acids had united with one another.

The products obtained were not natural proteins, however, even though they were made from amino acids. The special amino acids mentioned above contained either an extra amino or an extra acid group. In normal proteins, these extra groups do not take part in chain formation, but this had occurred in the heating process. Unnatural chains, even branched chains, had been produced. Further, some of the amino acids had been converted into their mirror-image forms, so that both types were present. Others had been converted to colored substances, pigments, which were also built into the chains. The term "proteinoid" rather than "protein" was applied to the product, because of these features which distinguished it from anything present in earthly biology.

The proteinoids were felt to be worthy of further study, as they had interesting properties of their own. For example, various preparations showed weak catalytic activity for a number of chemical reactions, though this activity was not remarkably greater than that possessed by the same amino acid mixture before it was heated. Much more striking, however, were the transformations exhibited by certain types of proteinoids when treated with warm water under appropriate conditions. One such treatment was to dissolve them in warm water and allow the solution to cool slowly. By this very simple process, the microspheres were obtained. Ten billion of them could be prepared from one gram (one-thirtieth of an ounce) of proteinoid.

We must note here that microspheres provide a fine illustration for the phrase "easy come, easy go." They can be dissolved quite readily be changing the acidity of the solution in which they were formed, or by adding additional water to that solution. It was this fragility which suggested the bubble analogy to me. If such steps are avoided, however, microspheres can be maintained and manipulated for considerable periods of time. Extensive and detailed studies have been carried out with them to explore their properties.

One immediate attention-catching feature is their resemblance to certain single-celled organisms, in both size and appearance. In cross section they look like bacteria, with internal compartments and double-layered outer boundaries that suggest membranes. Other preparations appear similar to ancient algae microfossils. Furthermore, microsphere preparations contain units fused together, a posture that suggests cell division. Fox himself has been cautious in interpreting this ability to divide in two: "This tendency is observed in soup droplets, mercury droplets, oil droplets and it occurs in molten glass droplets on the moon as well as in proteinoid microspheres."

In fact, while attending a recent meeting of the American Association for the Advancement of Science, I was struck by a lovely photograph of what I presumed to be proteinoid microspheres. Tiny circular microscopic shapes, some fused together, hung suspended against a transparent background. They were not proteinoids at all, however, but particles of volcanic ash from Mount St. Helens. They had been thrown into the air as molten lava and taken spherical shapes due to surface tension before they hardened. Crystallographer J. D. Bernal, noting the variety of shapes available in nature, has commented that the forms taken by the microspheres are not uncommon. "Any resemblance to organisms, such as the presence of double spheres indicating fission, is probably fortuitous," he concluded.

Sidney Fox and his collaborators would disagree. They feel that the many additional lifelike properties exhibited by the microspheres, taken together with their appearance, establish them as objects of importance. Lists have been prepared that include response to stains used for bacteria, catalytic activity, membranelike properties, electrical activity, sensitivity to light, and even reproduction. As a sample, we will consider this last property, quoting Fox directly: "The microspheres reproduce in a primitive manner, involving crystal-like growth. They do this through any of several processes: binary fission, the formation of budded microspheres followed by separation and growth of the separated buds, and also through what looks like sporulation and partition."

In the budding experiments, microspheres are placed in solutions saturated with proteinoid. As proteinoid accretes on their surface, they expand and form buds. Mechanical agitation of the suspension causes the buds to be split off. When the buds are placed in fresh protein they grow, "forming a second generation."

In summarizing this behavior, Professor Fox and associates like to compare the microspheres to primitive cells. In some statements they indicate a belief that microspheres could even evolve to present-day cells: "Accordingly, we believe that the proteinoid microsphere is capable of evolution to a contemporary cell, even though that capacity has not yet been fully demonstrated." At other times, however, they point out that it is only a model, a simulation, of a primitive cell. This ambiguity is covered by attaching the prefix "proto" (which is intended to mean minimal, incomplete, or primitive) to the properties they describe. Thus they write of protocells, protoorganisms, protoreproduction, protometabolism, protoevolution, and protosexuality.

Under some conditions, separated microspheres were made to fuse together, exchanging material. This phenomenon was considered relevant to "the origin of protosexuality in protocells," and in a later review it was called "a model of the origin of communication." Summarizing these and other demonstrations, Fox stated: "Proliferative protoorganism has been synthesized in the laboratory. Again, a full demonstration of total evolvability to a contemporary cell is yet to be made."

If this interpretation were to be accepted, the above accomplishments would in themselves reflect an enormous contribution to scientific understanding of the nature of the life process. However, the claimed significance was extended still further. Microsphere formation has been represented as the significant event involved in the actual formation of life on the early earth:

> The research of a number of investigators, especially space bioscientists, has suggested that the primitive Earth was a "rain forest" of organic compounds. It is now equally well established that the surface of the primitive Earth provided a rich lawn of varied macromolecules, especially thermal proteinoids. When these aggregated, on the touch of water, the latter became individuals poised for Darwinian selection.

These protocells, capable of primitive replications, would, on further evolution, develop the capacity to make authentic proteins as well as nucleic acids. The modern cell would gradually emerge, according to this account.

The geological plausibility of microsphere formation was a focus of much of the early criticism directed at the Fox group. Could temperatures as high as 150° to 180°C occur at all on the early earth, and if they did, would amino acids and other compounds survive extended exposure to them? Stanley Miller and Harold Urey came to a negative conclusion in a 1959 paper, stating: "It is difficult to see how the processes advocated by Fox could have been important in the synthesis of organic compounds."

Later Miller, in a book with Leslie Orgel, asked "whether there are places on the earth today with appropriate temperatures where we could drop, say, 10 grams of a mixture of amino acids, and obtain a significant yield of polypeptides.... We cannot think of a single such place."

The Second International Conference on the Origin of Life, the one held in Florida in 1963 at which Oparin first met Haldane, was organized by Sidney Fox. He and his colleagues presented data on their system in several papers at the conference. There was strong dissent concerning the possibility that such events could occur on the early earth. Carl Sagan, for example, stated: "I would like to see an order of magnitude calculation in which a probability is assigned to each scene of the scenario, and a total polypeptide abundance over geological time derived." (If only he had extended this request to the naked gene as well!) Geologist J. R. Vallentyne subsequently added, concerning Fox's prebiotic proposal: "Every time a geologist hears it, it pricks him under the skin and that's why you get such reaction to it from people who think in terms of earth history."

Professor Fox and his co-workers have been agile and flexible in attempting to meet such criticisms. Initially they suggested the rims of volcanoes as plausible sites where the necessary temperatures could be provided that were needed to form the proteinoids. Rain would subsequently remove them from the volcanoes and convert them to microspheres. To illustrate this concept, a sample of lava was collected at a volcanic site in Hawaii and brought to the laboratory of Professor Fox. The preparation of microspheres was then carried out within a depression of the lava sample.

This scenario was subsequently extended to other locations. It was found that lower temperatures (85 °C) could be used to make proteinoids, provided that the heating time was extended from hours to months. Thus, the hotter deserts of the earth joined the volcano rims as sites where the necessary heat could be provided.

When I interviewed him in 1983, Professor Fox mentioned another idea, one in keeping with current developments. He suggested that the hydrothermal vents at the bottom of the Pacific Ocean would provide the necessary temperatures rather nicely. Amino acids could form there and then be heated together to yield proteinoids. He paused for a moment when I asked how the necessary *dry* heating conditions would be found at the sea bottom. He then put forth the idea that a solid plug of amino acids might form in the mouth of one of the vents. As superheated water boiled off near this area, the amino acids would be converted to proteinoids. The plug of material would subsequently come loose. The entire procedure might not work more often than one time in ten, but that ratio of success should be sufficient.

Apart from the problem of location, other questions have been raised concerning the microsphere scenario. How were the necessary concentrations of amino acids assembled? Were the necessary special amino acids present in any quantity on the early earth? Would other chemicals that may have been present interfere with the process? Would not the microspheres, if formed, have dissolved on exposure to fresh water? Whatever the plausibility of the various steps, it must be noted that they are far simpler than the proposed prebiotic preparations of nucleic acids, which have received much less criticism.

I would propose that we put aside the question of the prebiotic plausibility of microspheres, for it is overshadowed by a much more significant one: Is it really true that a primitive cell with a number of lifelike properties, and ready to evolve, can be prepared in two simple steps (heat and add water) from a mixture, any mixture, of simple chemicals?

We have spent chapters in arguing why it is extremely unlikely that a structure with that degree of internal organization could arise by chance. Sidney Fox, in fact, would agree with this argument, since he proposes that the information needed for the construction of his protocell was already present in the original amino acid mixture. When the amino acids combined upon exposure to heat, they did not do so at random, but in a nonrandom manner, dictated by their individual chemical preferences. This process resulted in the formation of the protocell, which was ready for evolution by natural selection.

If this explanation were accepted, then much of the mystery of the origin of life would be removed. The nature of the essential steps would be understood. Whether this exact recipe had been followed, or some equivalent one more compatible with the actual conditions on the early earth, would be a matter of historical detail. The central principle would be grasped. But what exactly would this central principle be? There is more here than meets the eye.

To explore the question, perhaps it is best to put amino acids aside and recall our favorite animal, Charlie the Chimp, and his typewriter. We calculated earlier that if Charlie struck the keys at random, it was very likely that stars would have turned to ashes long before he produced the message "to be or not to be: that is the question." So let us de-randomize him. We then must ask how we will do this. There are an immense number of ways of departing from randomness.

We will introduce an element of reality and give Charlie a right-handed bias. Assume that he hits the right side of the keyboard somewhat more than the left side. Has this nonrandomness improved our prospects? No, they have become worse, as the majority of the characters needed for our message occur on the left side. Nonrandom is not necessarily better. On the other hand, prospects would improve if Charlie had a left-handed bias. But the odds are so steep against our desired message that it is still immensely probable that the stars would be gone before Charlie typed those words.

For this enterprise to succeed, Charlie's departure from randomness must be guided by some organizing force. Dr. Midas earlier played this role, when he stopped Charlie after each correct letter was typed. Even Dr. Midas would have been thwarted had Charlie typed so rapidly that he finished each line before the correctness of the first letter could be interpreted.

Amino acids are not abstractions; they will certainly combine with some measure of nonrandomness when heated together. The same result would be expected from any combination of real chemicals. But amino acids are also dumb, even more so than our chimpanzee. There is no obvious link between the condition that causes them to unite, dry heat, and the supposed product, a primitive but functional cell, any more than there is between the free-flying fingers of a chimp and our desired result, Shake-spearean prose. Yet some competent scientists believe that this event has clearly taken place, at least in the amino acid case. What is going on?

One way that I have approached this problem on my own is to ask what I would think if I saw a chimp approach a typewriter and type out a number of English lines, among which was the desired phrase from *Hamlet*. I would conclude that someone had rigged the typewriter and/or trained the chimp. In the chemical case, however, the corresponding elements are the amino acids and the chemical laws that govern the heating process. If these are sufficient to generate a primitive living cell, then someone has arranged the laws of chemistry so that they operate for our ben-efit.

We encountered this type of thinking in earlier chapters when we considered the response to the Miller-Urey experiment and the concept behind the design of prebiotic experiments. Here again we find the same assumptions, an example of what I choose to call predestinist thought. It is presumed that the rules that govern the connection of amino acids during a heating pro-cess will necessarily direct them into combinations with proper-ties useful to life. As we discussed earlier, such desirable results are not likely to be the result of good luck. An unstated, essen-tially religious assumption is made: the Creator has arranged things that way.

This assumption, of course, lies outside of science. Perhaps, if all other explanations should fail, in the end we will have no op-tion but to accept the idea of supernatural forces. Until we reach that point, however, we must look for rational ways of account-ing for the data.

One simple alternative is to presume that the properties of the microspheres are less significant than claimed. Suppose, for ex-ample, that our monkey had typed a sentence containing more numbers than letters, rather than a phrase from Shakespeare. It would be nonrandom but unimportant, indicating only that he had a preference for the upper part of the keyboard. Similarly, the various properties shown by the microspheres—division, weak catalytic activity, a double-layered border, electrical sig-

nals, and the rest—may be somewhat general properties of microscopic particles of a certain size and unrelated, or only slightly
related, to the actual processes of life.

During my childhood, I learned that I could make the shadow
of a dog with my hand. I needed only to point my thumb out,
bend in my index finger, and hold my hand before a light to
produce the image of a dog's head on the wall. I could enhance
the effect by moving my pinky while making barking noises.
But this form was not a dog, nor could it ever become one; it
was merely shadow play. In the same way, the properties of
the microspheres, while entertaining, may be merely shadow
play.

There has in fact been an entire history of tiny particles with
supposed lifelike properties, which is recalled in the origin-of-
life text of William Day. In 1892, for example, German biologist
Otto Bütschli rubbed droplets of olive oil with alkali and obtained tiny amoeba-like shapes that moved about and engulfed
particles. In the early years of this century, Stéphane Le Duc,
professor at the Medical School in Nantes, France, prepared
forms resembling algae and tiny mushrooms out of inorganic
chemicals. He called his efforts the new science of "synthetic biology." His followers escalated his claims, converting gelatin,
glycerol, and salt into "cells" said to have all the properties of
life. These transformations were accomplished by using the mysterious energy of newly discovered radium.

Such demonstrations continue to the present day, causing
some question whether Professor Fox's imitators may not do
more damage to his cause than his detractors do. At the 1983
meeting of the International Society for the Origin of Life in
Mainz, Germany, for example, two competing concerns were
present, each with their own poster demonstration.

An Indian group, headed by Krishna Bahadur and listing
twenty-six co-workers, displayed the virtues of the Jeewanu, mi-
crostructures named after a Sanskrit word meaning "particles of
life." Many photographs documented the cell-like appearance of
the Jeewanu (when stained they actually resembled microscopic
stuffed olives). They could be prepared by exposing any of a
wide variety of chemical mixtures to sunlight. One typical recipe

used mineral substances and formaldehyde. In addition to their cell-like appearance, the Jeewanu had "properties of growth from within, multiplication by budding," and "metabolic activities." Further, they showed enzymatic and photosynthetic activity, and were sensitive to antibiotics and sulfur drugs. They were termed "protocells."

A Japanese display at the same meeting advertised the properties of the marigranules. Like the microspheres, they were made from amino acids, using, however, a very different mixture than that employed by the Fox group. They could be prepared by strongly heating the amino acids in a liquid medium whose composition resembled sea water. No dry heat was needed. The marigranules had a suitable cell-like size and appearance, and were termed "models of organized particles produced in the primeval sea in the course of chemical evolution." Marigranules could be prepared by heating sugars, as well as amino acids. The chemical bonds within them had little relation to those present in life today.

One result of the intensifying competition may be an escalation of the claims made on behalf of the various "lifelike" preparations. A German group, several years ago, used an amino acid mixture similar to that used for microspheres but modified the recipe so that larger, fluorescent particles were obtained. Because of their size and shape, they were compared to eukaryotic cells. "They are about ten times larger and appear to develop more complex wall and sheath structures, tending to make up solid tissue-like formations." The German group further claimed that "different tissue cells can be distinguished in the grains." Their forms "resembled plant tissues." They concluded that "it seems credible that the processes demonstrated in our experiment were natural events on the young Earth."

An unfortunate comparison was made at this point: "The lumispheres resemble not only Fox's microspheres but also the microfossil Isuaphaera from the 3800-million-year-old Isua quartzite of Greenland." As we have seen, the supposedly yeast-like Isuasphaera appear to be ordinary mineral inclusions, and not a fossil at all.

In any event, this claim for the preparation of plant tissues from amino acids has been bettered by the Fox group, who have compared their microspheres to nerve cells. At one 1983 meeting, the electrical patterns obtained where electrodes were placed in a microsphere preparation were likened to the tracing of brain waves obtained from a sleeping monkey.

In my own interview with Sidney Fox he spoke with enthusiasm of this development, calling the phenomenon "excitability" and attributing it to the membrane of his protocell:

> In an artificial cell that has patterns of excitability which qualitatively and quantitatively are indistinguishable from what you get from brain neurons and other kinds of neurons, we're well on our way to understanding the origin of the mind. . . . The origin of life and the mind were synonymous. The roots of both life and mind appear to have been a pigmented thermal copolyamino acid membrane.

The latter phrase is the term Professor Fox used to describe preparations of the type he had been studying.

I hesitate to comment on this particular point lest I injure the feelings of any conscious microspheres. But something must be said on the situation as a whole, so we will call on the Skeptic for his point of view.

He notes that long lists of "lifelike" properties do not demonstrate that a system is alive or capable of life. Generally, those who prepared the systems have been careful to avoid such claims, using prefixes like "proto" or the term "model." But when such systems are readily made, what is really demonstrated is that their properties are of minor importance. The important ones are those that distinguish the model systems from truly living ones.

The situation has been mirrored in lotteries run by hamburger chains in the United States. To obtain a large prize, the winner must assemble a picture from perhaps nine fragments. A packet containing one fragment may be obtained simply by requesting it at any outlet of the hamburger chain. Those who enter the contest find that it is easy to assemble as many as eight of the fragments. Enthusiasm and excitement mount rapidly, and many visits to the hamburger outlets are made. But the last fragment is nowhere to be found. Other contestants find that they have the same experience.

In fact, it is the last fragment that controls the lottery. Whatever the number printed equals the number of prizes to be awarded. Possession of it alone essentially guarantees a win, while possession of the remainder is irrelevant.

What, then, is the missing fragment in the case of the origin of life, that which distinguishes a system capable of life from shadow play? It is the ability to grow, reproduce, and evolve. The system must convert simple materials in the environment to more of itself, not simply in the way that a rolling snowball gathers more snow, but in a manner that copies the internal organization of the system. Further, during this process, the system must increase its capabilities. Through evolution it must "learn" new functions which enhance its chances for continued survival and growth.

In the ideal experiment, an authentic primitive cell would be set up with an energy supply in a simple environment, and be allowed to grow and evolve continuously, without further intervention by the experimenter. Could the microspheres pass this test and demonstrate their special nature as a self-organized form of matter capable of further evolution into recognizable life?

I put this question to Sidney Fox, but he disagreed about the type of experiment that should be run. He felt that evolutionary progress required a *"stepwise changing environment,"* with the scientist specifying the changes. "I don't expect to do experiments that will run themselves after I leave them. At the least, I must feed my microspheres. Later in evolution, if a mother abandons its baby it dies."

At this point our philosophies part from one another. The critical issue does not concern whether the environment is static or has a preset cycle of some kind (between wet and dry, for example, or hot and cold). Rather, it involves the choices to be made when things go badly. Shall the investigator, acting as a caring parent, intervene continuously to ensure the survival of the system rather than accept a negative result? If he does so, he might demonstrate his ingenuity, but he would also play the role of the organizing agent. No self-organization by the system would be demonstrated. Only when negative results are permitted and theories can be abandoned will science progress in this area.

We must not prejudge the microspheres, however. Perhaps some demonstration can be made that would satisfy both Professor Fox and his critics. If so, then he will certainly have the last laugh. It would be against the spirit of science to declare flatly that it is impossible.

To presume that such circumstances *must* exist is also unscientific, however. Such a presumption would establish the study of microspheres, Jeewanu, or other such particles as yet another area of mythology, outside the possibility of negation. For the reasons stated earlier, it seems unlikely that such a demonstration will ever be made. Any system with sophisticated abilities such as those attributed to the microspheres is likely itself to be the result of an extended evolutionary sequence, rather than a one- or two-step process. It matters little to this conclusion whether this process is called spontaneous generation or self-assembly.

If we accept this reasoning, then the missing fragment in our picture of the origin of life would be a principle that governs the gradual evolution of simple chemical systems into more sophisticated ones capable of replication and Darwinian natural selection. The search for this principle has begun.

Ripples

In my years of laboratory training, I learned that only a few types of behavior might commonly be expected when chemicals were mixed together. Quite often, the result was the dullest imaginable. Nothing visible happened. It might take hours or days of work for me to discover whether anything important had in fact taken place, but there was nothing that my senses could detect.

Occasionally, I was rewarded by a signal that something was going on. Bubbles of gas might appear in a liquid, as we see when we open and pour out a bottle of club soda. (Too much of this would be very undesirable, as it would lead to the chemist's bugaboo—an explosion.) Alternatively, a solid might form suddenly. This could be very entertaining. One of the early factors that propelled me to the study of chemistry was the sensual experience of seeing an immediate, lush, bright yellow solid formed when two colorless liquids were mixed together.

Much more common, but far less satisfactory, was the gradual appearance of a dark, sticky tar when new combinations of organic chemicals were heated together. Such tars usually signaled the failure of the intended reaction. My failure in such an experiment was accompanied by an additional punishment: I had to find some way of cleaning the glassware which contained the gunky mess. After a series of such disasters, I would sometimes try to cheer myself up by rerunning a particularly nice known reaction. It was one that, after a night of standing unattended, deposited a flaskful of long, large, shiny, and incredibly beautiful crystals.

All of these differing forms of chemical behavior had one feature in common, which was too obvious for me to notice at the time. They went about doing whatever it was they were going to do, and when it was over, they stopped. They did not shift back and forth, like the lead in a horse race, or reverse themselves entirely, like the tides.

Recently, however, a very different type of chemical reaction has attracted attention. In a limited number of known reactions, oscillations are set up. The amounts of certain chemicals in the mixture rise and fall periodically as the reaction proceeds. If the components are selected cleverly, this can be made apparent visually, with a startling effect. The reaction may change from colorless, to gold, to blue, to colorless again, repeatedly.

In some cases, beautiful spatial structures, such as ripples or spirals, may appear. One such reaction was described in *Scientific American:* "... the solution is initially a uniform purple.... As the reaction proceeds white dots appear and grow into rings and sets of concentric rings which annihilate each other when they collide. One observer compared the emergence of white dots against the purple background to seeing stars come out."

Some fraction of these systems also can show a related behavior, termed chemical chaos. Oscillations occur, but they are not periodic. Rather, they increase and decrease in a seemingly random and unpredictable manner.

These reactions have been studied, but are not understood fully. Although only a few simple chemicals may be required to set up such a system, chemists cannot yet predict entirely new combinations which will show behavior of this type, or suggest what new types of behavior will be possible. There is more than enough motivation to continue the studies, however, as a number of scientists feel intuitively that they will provide the missing clue needed to understand the origin of life.

We have seen that self-replicating systems capable of Darwinian evolution appear too complex to have arisen suddenly from a prebiotic soup. This conclusion applies both to nucleic acid systems and to hypothetical protein-based genetic systems. Another evolutionary principle is therefore needed to take us across the gap from mixtures of simple natural chemicals to the first effective replicator. This principle has not yet been described in detail or demonstrated, but it is anticipated, and given names such as chemical evolution and self-organization of matter. The existence of the principle is taken for granted in the philosophy of dialectical materialism, as applied to the origin of life by Alexander Oparin.

Creationists, as we shall see, have equal faith that no such principle exists. They hold that the exquisite organization that we observe even in the simplest living things is not the result of an evolutionary process but was generated in its present form by an even more organized and perfect Creator. In the words of their spokesman Henry Morris: "The creation model postulates a primeval creation which was both complete and perfect, as well as purposeful." Since that time things have run downhill rather than uphill, according to a scientific principle called the second law of thermodynamics.

According to Morris, if random matter has really evolved from chemical elements to man, "then there must obviously be some powerful and pervasive principle which impels systems toward higher and higher levels of complexity." Morris, however, denies the existence of this "basic law of increasing organization."

Sir Fred Hoyle has taken up essentially the same theme, and called for proof or disproof by experiment:

> If there were some deep principle that drove organic systems toward living systems, the operation of the principle should easily be demonstrable in a test tube in half a morning. Needless to say, no such demonstration has ever been given. Nothing happens when organic materials are subjected to the usual prescriptions of showers of electrical sparks or drenched in ultraviolet light, except the eventual production of a tarry sludge.

An earlier exponent of this theme was William Jennings Bryan, the noted politician and opponent of evolution. He stated: "If there were in nature a progressive force, an eternal urge, chemistry would find it. But it is not there."

Certainly, the demonstration called for by Hoyle has not been performed. There would be little reason to expect success in such an effort, if it were necessary to overturn the second law of thermodynamics in the course of the experiment. But is this really the case?

We mentioned the first law of thermodynamics briefly in an earlier chapter. It stated that energy cannot be created or destroyed, but can be converted from one form to another. The second law elaborates on the rules that govern these conversions. It specifies the processes and conversions that can occur, and those that are impossible. A key concept is a term called entropy, which we can equate with randomness or disorder. The second law states that entropy will increase in any spontaneous process that involves the entire universe, or some part of it that is sealed off from the remainder (such a part is called a closed system). Thus, in such processes that take place of their own accord, things do not become organized, but rather they get disorganized; entropy increases.

We can understand this intuitively on the basis of our daily experience. If we put a drop of ink into a glass of water, the color spreads out until it is distributed evenly. We will never see this process run backward, with the color recollecting itself into a single drop. Similarly, if a hot object and a cold one are put into contact with one another, they will eventually come to the same temperature. The average motion of the molecules in the two objects (which we experience as heat) will be equalized.

The idea of disorder is closely linked to considerations of probability. If we assume that every molecule of the ink pigment in a glass of water has an equal chance of being in the upper or lower half of the water, and come to its "decision" independently of the other molecules, then there is a finite possibility that the ink will all collect in the upper part of the water, leaving the lower part colorless. Yet the odds against this happening can best be described by noting that this event would take us to floor 100,000,000,000,000,000,000 of our Tower of Numbers! We need not sit around waiting for it to occur.

At first glance, living things appear to be in a horrid state of improbability, greatly in violation of the Second Law of Thermodynamics. Take the amino acid units in the enzymes of our bodies, for example. The two mirror-image forms of each amino acid have the same chemical energy, and are equally likely to occur. We would expect a random collection of amino acids to contain half of the D-form and half of the L-form. Yet the amino acids in our enzymes are essentially all in the L-form. This improbability is comparable to that of the ink-and-water example discussed above. We could pile up improbabilities by considering the other forms of organization present in our cells, but there is no need to do so to make our point.

Can this state of affairs be reconciled with the second law? The answer is yes, rather easily. Living things do not exist as a closed system, isolated from their environment. When isolated, they die. To make things concrete, let us imagine a few bacteria placed in an environment which contains some simple organic compound as a food source, needed inorganic salts, and an oxygen supply. We will seal off this total system, so nothing can get in or out. According to the Second Law, the entropy of the contents of our sealed box must increase. Yet our bacteria would multiply happily for a time, converting the simple chemicals into L-form amino acids and ultimately into many more bacteria. This conversion would be accompanied by a decrease in entropy of the chemicals involved.

No contradiction is involved in these events, as we have not yet described the entire situation. In the process of growth, the bacteria combined a portion of the organic material with oxygen, producing carbon dioxide and water. This conversion was accompanied by a great increase in entropy, more than enough to offset the decrease involved in the creation of new bacteria. Thus the entropy of our entire container went up, in accordance with the Second Law.

Ultimately, when the food supplies in their limited environment ran out, the bacteria would cease to grow and would die. They could be kept alive if we opened the box, increasing the environmental resources available to them. The bacteria of this planet, and the rest of us as well, can decrease our entropy continually, as these changes are balanced out by greater entropy increases within the sun, the ultimate source of sustenance for almost all life on earth.

It is difficult to keep track of entropy changes within the sun, so another set of terms is used to express the same relationship. We say that life receives a supply of free (meaning available) energy from the sun, and uses this energy to maintain and increase its state of organization. In the same sense, the bacteria in the box used the energy released by the reaction of their food supply with oxygen to sustain themselves.

There is a vital message in this story. Improbabilities that are hopeless in terms of random events, such as the formation of only L-form amino acids from simple chemicals, can readily be achieved if a suitable energy supply is available. Nor is the cost prohibitive. Items carrying an improbability that would place them thousands or millions of floors up on our Tower of Numbers can be purchased for the energy stored in a pinch of ATP.

The energy released from the sun, then, is the source of the improbability present in life today, and was also the driving force for the organizational steps involved in the creation of life. Chance is an incredibly less effective tool for these ends. But the crucial problem remains. How was the sun's energy first harnessed for this purpose? Available energy is not normally converted to chemical improbability. Normally, much or all of it transforms to heat, and becomes unavailable. Sunlight irradiating a junkyard will not cause the junk to assemble into a Boeing

747, it will merely give us warm junk. Bacteria maintain an intricate machinery to achieve the transformation of chemicals and energy into more bacteria. Let us quote Henry Morris again: "The question is not whether there is enough energy from the sun to sustain the evolutionary process; the question is *how* does the sun's energy sustain evolution? ... Where does one see a marvelous motor which converts the continual flow of solar radiant energy bathing the earth into the work of building chemical elements into replicating cellular systems?"

Yet many scientists remain convinced that such a marvelous motor exists. As stated by John Keosian: "Matter driven by energy in an open system can go on to higher and higher levels of organization." As we said, this is accepted as an article of faith in dialectical materialism, which presumes that the organizing process extends beyond atoms, microbes, and men to the development of higher societies.

The issue concerning the origin of life is better approached through mathematics than politics, however. Ilya Prigogine received the Nobel Prize in chemistry for his development of the thermodynamics of nonequilibrium states. He has stated that "a prebiological system may evolve through a whole succession of transitions leading to a hierarchy of more and more complex and organized states." During this development a series of instabilities, called "dissipative structures," emerges.

Others, such as Manfred Eigen and physicist Harold Morowitz, have also attempted to approach such situations through calculations. We encountered Eigen's hypercycles earlier. Morowitz has also concluded that material cycles would be involved in chemical evolution, and that "the organizing principles of molecular chemistry appear sufficient to direct systems along highly specific pathways toward living forms."

They may be sufficient, but we can't be sure what they are. Morowitz has added: "There may be other principles still undiscovered that are important to the development of highly ordered prebiotic systems." The mathematics reassures us that a solution exists, but it is not *the* solution. We need to see it demonstrated in the laboratory, to watch a system evolve in stages in accord with the concepts of chemical evolution.

The ripples and rings that we described earlier represent the first step on a path of this type. They demonstrate how simple chemical mixtures can produce organized structures. Our need is to find a system that keeps on going, one in which the structures get more and more complex, the chemical cycles more organized. The rings and colors are well-chosen symbols, but not the most important feature. We wish to find an evolving chemical system that ultimately produces a replicator. A different type of prebiotic experiment would be needed to identify the needed conditions.

Imagine a mixture of amino acids and other chemicals, such as the one produced in a Miller-Urey experiment. Unlike the Miller-Urey case, however, the energy source is not turned off after a week but continues to bombard the mixture. The chemicals continually break up and recombine to form new mixtures.

Presume now that one chemical (or set of chemicals) interacts with the energy source in a way that stabilizes it, and increases its extent in the mixture. A new combination would arise, which would in turn interact with the energy source, perhaps further favoring certain chemicals at the expense of others.

A start has been made in developing such a system. When amino acids are subjected to alternating cool, wet and then warm, dry treatments on a clay-mineral surface, they unite to form short chains and come apart again. One short peptide chain, made of only two amino acids, has been shown to favor the joining process.

If we were to speculate about possible further developments, we could visualize an evolving community of short peptides interacting in some manner with an energy source, such as the sun. Collectively, they would work to favor the formation of amino acids in preference to other chemicals, the union of amino acids to peptides, and the preparation of useful peptides rather than functionless ones.

Direct replication would not yet have developed, and heredity would be carried by the ensemble as a whole. Let us suppose, for example, that the community had a need to break down a type of tar that occurred in its environment. It would not be able to prepare a single specific enzyme to do the job effectively, but rather would make a range of molecules which had the needed ability to a greater or lesser extent. When one very fit molecule was destroyed by chance, its functions would be performed by the next-best molecules available until another "expert" one came along.

The situation has some resemblance to the way in which particular tasks are handled in our society. We cannot replicate our best brain surgeons and violinists directly. When they retire or die, others take over, and more are trained continually.

According to our speculation, an evolving amino acid and peptide system of this type would slowly gain in complexity. If the mathematics that describes these processes is correct, progress would be made in a series of fits and spurts, rather than in a smooth manner. However, no steps of great improbability would be involved, once the process was under way. Eventually, as more complex molecules arose, selection pressure would develop to favor the emergence of a system of enzymes capable of direct replication. We would have arrived at the stage where Darwinian natural selection could take over.

The Skeptic must have a hearing at this point. Ironically, his viewpoint will be close to that of a Creationist. He will remind us that the huge gap between the ripples and the replicator has been filled only by calculation and conjecture, but not by experiment; that chemical evolution has been affirmed and denied, but not demonstrated. We still have no way in which to predict the type of mixture that would be a suitable starting point. To illustrate the process I have used chemicals familiar to us, amino acids.

Amino acids, when combined, do show the right kind of properties. Further, they are prevalent in life today. For these reasons, they are logical ingredients for a system of chemical evolution. It is not clear, however, whether conditions on the early earth were suitable for forming them and concentrating them. As we have seen, there are also problems involved in combining them by splitting out water to form long chains.

For these reasons, an alternative solution has been suggested. Perhaps the processes of chemical evolution, and even the early steps of natural selection, used other chemicals: ones that were surely present on the early earth, but are no longer needed or used in our life today.

Mud

A playwright starts with an empty stage and then specifies the set and props needed to carry out his drama. Quite often, origin-of-life scientists function in the same way. The early earth is taken as a featureless backdrop, to be decorated for the benefit of their theory. They call for a reducing atmosphere, or a rich supply of meteorites, high or low temperatures, and all manner of specific chemicals, pleading that their requests are reasonable or, in any event, needed for the success of the drama.

Well-designed sets may work wonders in the theater, but the opposite spirit prevails in science. The theory that works with the least number of arbitrary assumptions will be the most satisfying one. We do not know whether the early earth had a particular atmosphere or whether organic compounds were plentiful. But land was there, as were hills or mountains, also some form of atmosphere, and wind. Water was present, and therefore rain, rivers, and seas.

The action of wind and water on the mountains would produce boulders and rocks. By further weathering, sand, silt, and even finer particles, clay, would be produced. Water would mix with these substances, forming mud, which would be carried downstream by rivers. When the rivers slowed in flat places, the mud would be deposited as sediment. As sediments piled atop one another, the lower sediments would be compacted into new rocks. At a later stage, uplifted by geological forces, these rocks would in turn be subject to erosion.

Rocks are made of chemicals and are transformed by reactions, as well as by physical forces such as breakage. Water, percolating through rock pores, dissolves some of the minerals present and transforms others, leaving altered substances. New rocks may be formed not only by compacting old ones, but also by deposition from evaporating water solutions. Minerals not only take part in various processes but affect their course, acting as catalysts. They do so most effectively when they are divided into the smallest particles, those with the largest surface area, the clays.

As clays were surely onstage and active in the early days, some investigators have tried to involve them in the plot of the orgin of life. Crystallographer J. D. Bernal postulated that they helped assemble the molecules important to biology, by collecting them from the seas in which they were dispersed. Others have given them an important supporting role, catalyzing the preparation of these molecules before the start of life. We have seen how the action of some unknown magic mineral has been held as a reservoir of hope, a possible *deus ex machina* that would resolve the unsettled drama of prebiotic nucleoside synthesis. However, in efforts to reenact this drama, modern clays have been unwilling to play this role to any important extent.

The most exciting suggestion about the part played by clays in early life comes from chemist Graham Cairns-Smith. He gives them the leading role. They did not simply assist the organic chemicals, but were the living beings themselves: "For a picture of first life do not think about cells, think instead about a kind of mud, an assemblage of clays actively crystallizing from solution."

For most of us, rocks, however finely divided, have not been the substance of life but rather the clearest symbol of the opposite: nonliving matter. Sandy deserts and the barren surface of the moon come readily to mind. Biology texts often compare the properties of an animal and a rock in an effort to define life. When Pet Rocks were offered for sale some years ago, this was intended as comedy. Nothing about rocks, or mud, suggests that they are suitable construction materials for life. But we must remember that a bucket of tar also gives little clue to the wondrous biochemistry that can be developed with carbon compounds.

Let us pause and think of one familiar behavior of a common mineral, table salt. Imagine that we add salt to a flask of water until no more will dissolve. A single salt crystal is suspended in the clear solution, which is left open to the air. As water evaporates, additional salt will deposit on the surface of the crystal, causing it to grow. If some shock or imperfection caused the enlarged crystal to split in two, we could claim that the initial crystal had replicated itself.

This behavior has been used to challenge definitions of life that featured only growth and replication. In fact, the growing salt crystal can do nothing more. No further lifelike properties will emerge with time. But this feature need not be inherent in all minerals. Salt, whose chemical name is sodium chloride, is a very boring substance, composed as it is of only two types of atoms in a one-to-one ratio.

We can visualize its structure by drawing an extended tic-tac-toe grid of crossed lines and filling in alternating *x* and *o* symbols, so that each *x* is surrounded by four *o*'s, and vice versa. Let us extend this arrangement in our minds into a third dimension, so that each symbol is adjacent to six of the opposite kind. If we now replace *x* and *o* by sodium and chloride, we have the crystal structure of salt. However, this structure no more suggests the possibilities of mineral chemistry than the structure of diamond (a perfectly regular three-dimensional network of carbon atoms) reflects the great versatility of organic chemistry.

Other mineral structures offer far more promise than salt, particularly those made with oxygen and silicon, the two most common elements in the crust of the earth, as key ingredients. Just as one great vineyard may adjoin another famous one in celebrated French wine areas such as Bordeaux and Burgundy, so does silicon lie adjacent to carbon in the classification table of the elements used by chemists. Silicon, like carbon, prefers to bond to four other atoms at a time, a property that can lead to great chemical complexity. Silicon is unlike carbon in that it prefers a single type of bonding companion, oxygen, under earthly conditions. The silicate group, which we described earlier as a unit containing one silicon and four oxygen atoms, is the most common building block of rocks on earth.

The word "silicate" does not complete our chemical description, however. Each oxygen atom must choose a bonding partner in addition to silicon, and the manner in which it does this determines whether we get a substance not much more complex than salt or a system of great versatility, one perhaps capable of life.

More interesting possibilities arise when one or more oxygens in each silicate unit also join to another silicon atom, thus interconnecting multiple silicate units. Such connections can proceed linearly to form chains, in two dimensions to afford layers or sheets of silicates, or in three dimensions to provide frameworks. Those oxygens that do not bind to two silicons may combine with various metals, affording a variety of structures. Additional variety can be obtained by the replacement of occasional silicons by other appropriate atoms, such as aluminum.

Network silicates, bearing names such as quartz and feldspar, are the principal components of the volcanic rocks of the earth. The sheets, however, will draw our attention, because of their possible life-bearing properties. They can be attractive substances. In some cases, the layered arrangement of atoms is even reflected at the visible level. As a child, I would often walk in a park near my home and search for shiny mica crystals. I would peel off very thin layers and marvel at their transparency to light. Each visible layer of course contained enormous numbers of silicate layers.

Micas are formed directly when volcanic lavas cool under the proper conditions. When they weather, they are transformed by water to afford a group of related layered silicates called clay minerals. This use of the word "clay" differs from the one we used previously, which referred to a tiny size of particle that may be formed from any mineral. Clay minerals can of course be weathered to form clay-sized particles.

A variety of types of clay minerals can be found on the earth, the most common of which is called kaolinite. This substance is the principal ingredient in a mixture, kaolin clay, which is used in the manufacture of porcelain and other pottery. The name is taken from a mountain in China, Kao-ling, used as the source of the first such clay sent to Europe.

When viewed under the microscope, kaolinite crystals often appear as booklike aggregates of flakes. The stacks of "pages" may pile up to a much greater extent than pages in books, however. The stacks, viewed on edge, are not always straight but sometimes bend into wormlike or "vermiform" shapes.

Kaolinite crystals, like salt crystals, can grow by the accretion of additional material from solution. In one form of growth, new layers or "pages" are added to the booklike stacks. Because of such properties, clay minerals in general and kaolinite in particular are the favorite candidates of Graham Cairns-Smith as the building stuff of his living mineral system, his "vital clay."

If the sheets of clay minerals were always the same, the equivalent of blank pages, they would hold little interest for our purposes. Imperfections may occur in mineral structure, however, which mimic the role of printing on a page. Such defects can form in various ways—for example, by the substitution of other atoms for occasional silicons, or the fusion of two "pages" at an angle to form a bent "page." These structures are well documented, but the remainder of our story is made up of little evidence, and much speculation.

To continue, we must presume that new layers can be added to a growing silicate stack in a manner that copies the defects in the existing sheets of the stack. In our book analogy, the situation would correspond to the addition of extra photocopies of some document to a stack of such copies. Chemical circumstances exist in which the replication of a pattern of imperfections in a clay-mineral sheet could take place, and a German chemist, Armin Weiss, has reported preliminary experiments which document such an event. In his studies, the new sheets were formed within the stack, rather than at either end.

To mimic biological replication, we would want not only to lengthen the stack, but also to divide it into several stacks. Weiss has shown that this process occurs when the salt concentration in the water bathing the mineral is lowered. In nature this could occur when a rainy spell followed a relatively dry period.

A pattern of structural imperfections in a layer of clay may then serve as a two-dimensional mineral analogy of biological information storage in the base sequence of DNA. To complete the analogy to biology, it would be necessary that the defect pattern also affect the physical and chemical properties of the clay. This appears to be the case in some situations at least. The catalytic ability of clay minerals for certain organic reactions varies with the degree of aluminum for silicon substitution in the sheets, according to Weiss.

We are ready now to raise the curtain on Act One of our clay play, with the stage set on the early earth. We see various growing clay minerals in a suitable environment, say a porous stone permeated by flowing water containing dissolved minerals. The different species will compete with one another for dissolved mineral "food," with the winner replicating the most rapidly. We described this situation in Thursday's Tale in the Prologue.

We must now introduce another supposition. The replication of clay patterns is not fully accurate. Errors occur, sometimes resulting in the generation of clays with improved properties for survival and propagation. Natural selection among clay mineral systems has begun.

In this scenario, an evolving clay-mineral system was the first life to appear on earth. The plot offers many advantages not possessed by those featuring original organisms built of carbon. No special atmosphere or soup need be specified, only geological cycles of the type that still function on earth. Energy is provided by the forces that raise mountains from the earth, and weather them down again. No great leap in organization is needed to bring us from the initial chemical state to the first replicator. They are closely related to each other. Further, biological replication of our type requires the union of small molecules with release of water, an unfavorable process in terms of energy. Crystal growth, on the other hand, is a favorable process in many circumstances.

Let us continue with the first act. Mineral beings have evolved to some extent. What would they be like? According to Cairns-Smith, they would live at first in stable, protected environments, below the ground or near the bottom of the sea. Only later would they spread to more variable, exposed locations, close to the surface. They would be rooted in their location, like plants rather than animals in this respect. They would spread by taking advantage of the flow of streams and currents, as they fragmented during replication.

Minerals may not have the ability for exact molecular control of reactions that our own enzymes possess. Rather, they would influence the course of chemistry by the use of apparatus, much as a modern chemist does in the laboratory, but on a microscopic level. Various tubes, pores, membranes, pipes, and even pumps would be encoded in the information stored by crystal genes. Other minerals as well as layered clays would be used in their construction. The situation resembles that present in our biology today, in which instructions written in DNA are given final form using a variety of building materials. At the close of Act One, the earth might contain numerous communities of evolved organisms, each member of which, in the words of Cairns-Smith, might be compared to "a cardhouse with rooms of just certain sizes interconnected with each other in definite ways."

As Act Two unfolds, clay communities have populated the surface. This move affords them broader opportunities for dispersal. For example, a "vital mud" might dry to a fluffy powder and be scattered to new locations by the wind. As the capacities of mineral communities expanded, they would begin to experiment with new building materials, in particular organic molecules.

We need not presume that they intruded into any rich prebiotic soup. They could use solar energy and carbon dioxide in the environment to prepare organic substances by photosynthesis. The catalyst needed for such a process may either have been present in the mineral environment or arisen as the result of an evolutionary process.

We may imagine that early experiments in organic chemistry conducted by clay organisms ended with results much like those that I and other modern chemists have encountered: an apparatus coated with intractible tar. There may have been many evolutionary disasters before clays "learned" to control organic reactions. With time and experience, this capacity was developed.

The first organic molecules introduced into clay organisms may have played minor parts. They were used to alter clay consistency, sequester rare minerals, and serve as building materials. As they advanced in sophistication, they assumed more roles in mineral life.

As Act Two concludes, we will use Cairns-Smith's words to describe an active clay-organic community inhabiting an analog of Darwin's "warm little pond": "Think of Darwin's pond as an ecological system consisting of a community of well evolved clay-based organisms living in shallow water exposed to the sunlight." One member of this community would carry out photosynthesis using carbon dioxide, another would convert nitrogen in the atmosphere to a more usable form, another would collect rare minerals, and so forth. The Golden Age of mineral life was at hand.

In Act Three, I have modified the basic story of Cairns-Smith to better fit the theme of this book. Much time has passed, and enzymes have been devised. Their function of chemical control has proved better than the elaborate clay apparatus, and the pumps and vesicles have gradually been discarded. Genes of clay and their protein products have been encased in membranes, to gain protection and mobility. At some point, it becomes more efficient for control purposes to store genetic information and capacity in the enzymes as well, rather then keep them only in clay.

Only one step remains to complete the drama. Such dual clay-organic genetic system organisms would still be dependent upon a supply of dissolved silicates to replicate. Some organisms could free themselves of this restriction simply by discarding the clay genetic apparatus, retaining the alternative one. The transfer of control from minerals to carbon compounds would be complete, and thus the clay play would end. Modern evolution could begin.

The name that describes this drama is genetic takeover, the title used for the recent technical book by Graham Cairns-Smith. He has compared the process to the revolution in modern electronics, where compact and efficient solid-state instruments have replaced the tubes and wires of older devices.

Those origin-of-life scientists who attended the 1983 conference in Mainz, Germany, had a chance to observe another analogy. Johannes Gutenberg had developed his printing press in that city in the mid-fifteenth century. An excellent museum documents that event, and the general history of books from earliest times to the present day.

I was struck with the occurrence of a number of technological revolutions, of takeovers, in the history of written and printed information. The one made famous by Gutenberg, and still in use today, came quite late in the development of the art. The invention of writing itself, and of paper, preceded it. The printing press, of course, caused a revolution in the rate at which information was transmitted.

Similarly, there may have been a number of takeovers in the

development of the mechanism of biological storage of information, with earlier clay, protein, and RNA-based systems giving way to DNA late in the process.

The above play, a speculative account of the origin of life on earth, has many satisfying features and, of course, a major handicap. Before we can consider our evolution by genetic takeover from clay organisms, we must accept that such beings can exist on their own. A considerable number of scientists, for whom I coined the term "carbaquists" in an earlier book, are convinced that only a system based on carbon chemistry, operating in a water medium, can sustain life. Extreme carbaquists would limit the possibility of life to a nucleic acid and protein system similar to our own.

The demonstration of life forms made of minerals would overturn this point of view, and greatly expand our conceptions of the forms in which life in the universe may exist. This development would provoke a revolution in our understanding of the nature of life, and constitute an overwhelming achievement, even if it had nothing to do with the origin of our own type of life on earth.

A discovery of this magnitude will require considerable proof before it can be accepted. Fortunately, there are a number of ways to test the proposition. Experiments of the type run by Armin Weiss may be difficult technically, but fortunately do not require a trip to the bottom of the sea or to another planet. The ability of clays to store and express information, replicate, and mutate must be demonstrated rigorously and reproducibly.

Cairns-Smith has suggested a different and dramatic type of experiment—a mineral version of the test-tube evolution study with $Q\beta$ RNA. It would employ a continuous crystallizer, an apparatus already in use for other purposes. A supersaturated solution of minerals would flow into such a device, crystal formation and growth would take place within it, and a suspension of crystals would flow out the other end.

Let us suppose that two different species of mineral crystals form within the device. The first grows quickly but does not break up. Eventually, it washes out the exit pipe and is lost. The second type, however, not only grows but fragments rapidly. The new crystals that are formed compensate for the losses. If some variant formed by a random alteration can proliferate more rapidly, however, it will take over the chamber, as $Q\beta$ mutants did in Spiegelman's experiment. Studies in the crystallizer will tell us much about the possibilities of crystal evolution and the types of mineral most adept at it.

The best proof of the possibility of clay life would be its detection on earth today. Nothing in our clay play ruled out the survival of the original forms. The mixed crystal-organic hybrids that preceded organic life may have been devoured by their descendants, and remain only in fossil form today. Earlier versions made of clay alone would not compete with modern life for the same resources or environments, and may survive to this time. Even if they had perished due to geological changes, they could be expected to start and evolve all over again, because of the relative simplicity of the process.

Graham Cairns-Smith was cautious about these possibilities when I spoke to him at the Mainz meeting, despite the fact that we stood in a room surrounded by Roman artifacts and sipped wine to increase the courage that scientists need for unrestrained speculation. He felt that mineral life forms would be fragile and vulnerable to extinction. Only restarts, not original survivors, might be expected today. In searching for them we should seek out rare and unusual forms, such as elaborate kaolinite vermiforms, that dominated their environment. Alternatively, we might look for crystals in unexpected places, far from the source of the materials that formed them. He offered no immediate candidates, though he has at other times wondered whether some known vermiforms were not the result of a natural selection process.

We must wait for any final answer concerning clay life. If in fact we began in this way, it would be one of the most satisfying scientific answers. We inhabit this planet, and use its resources. Our bodies are placed into the earth when we pass on. How fitting if we were ultimately born of this soil as well, as suggested in Genesis 2:6-7: "But there went up a mist from the earth, and watered the whole face of the ground. And the Lord God formed man of the dust of the ground, and breathed into his nostrils the breath of life; and man became a living soul."

The various speculations in this chapter differ widely, but they share with one another, and with many other theories, the presumption that life did originate on earth. This supposition need not be true. The lack of firm evidence favoring the start of life here has caused some noted scientists to turn their thoughts elsewhere. We will consider their ideas next.

nine

The Comets Are Coming: Science as Religion

"There are those who believe that life here began out there." A phrase of this type was repeated at the start of each episode of a recent television space-opera series. As it was spoken, the screen showed a fleet of spacecraft headed for the planet Earth, in a massive galactic Exodus. The scope of these events, and the backdrop of star-studded space, sent a clear message. Our presence on this planet was not merely the result of some local accident, but rather had cosmic importance, affecting the entire galaxy.

The heavens on a bright night are a magnificent sight. I find it almost impossible to look up at them and not be overwhelmed by their majesty. I was spared this experience for much of my childhood, as I grew up under the hazy, glare-filled skies of New York City. Only occasionally, when my family was on vacation in the Catskills and I was allowed to stay up unusually late on a summer evening, could I enjoy the full experience. More often, I saw it in simulation, under the artificial sky of the Hayden Planetarium. Whatever the circumstances, once I had seen the effect I could understand the emotions of those who wished to move our origins out into the cosmos. They were similar to those of a chambermaid in a fairy tale, who secretly hoped that she had been born a princess and that someday her true identity would be revealed.

Such thoughts have come up repeatedly throughout history. It is not surprising that they reemerged as soon as the Oparin-Haldane hypothesis showed some weakness. When it began to appear likely that the early earth did not have the strongly reducing atmosphere required by the theory, several responses became possible. One was simply to modify or abandon the theory, but this option was not as attractive to some thinkers as an alternative one: abandon the earth, and move the origin of life elsewhere.

We will not consider here the popular fantasies in which ancient astronauts or earless extraterrestrials, our parents or at least our cousins, are lurking just around the corner, teasing us from time to time with a glimpse of their spaceships. The evidence for such events simply does not pass muster. Unless some hard data to the contrary should appear, we will prefer to remain with the simplest assumption: no external intelligence has interfered with events on this globe during the time covered by the geological record.

This record, of course, does not extend back to the origin of life, and we cannot exclude the possibility that the first organisms arrived here from elsewhere. They may have arrived by accident, or as a result of the efforts of conscious beings.

One theory of this type was publicized early in this century by a celebrated Swedish chemist, Svante Arrhenius, who won the Nobel Prize in that field in 1903. Arrhenius favored offbeat ideas. His Ph.D. thesis at Uppsala had correctly described the behavior of salts when they dissolved in water. This theory was received with little enthusiasm. He received the lowest possible passing grade for it, but his ideas were later vindicated.

This experience undoubtedly gave him the courage to advance a radical hypothesis about the origin of life: the theory of panspermia. Arrhenius proposed that small microorganisms were ejected from the atmospheres of life-bearing planets elsewhere in the galaxy. These microbes drifted through interstellar space as spores, propelled by the pressure of radiation from stars. One survivor of this process reached earth, to start life here.

This theory is discredited, in the opinion of most scientists today, although occasional dissent can be heard. Astronomer Carl Sagan and others have argued that the arrival of even one spore by this path, during the entire history of the universe, is an unlikely event. Further, any microorganisms of the type we know on earth would be killed by the radiation hazards, cold, and vacuum of outer space.

Such hazards would be eliminated of course if the microbes arrived as passengers in a suitable vessel. In the 1960s, Cornell University's Thomas Gold suggested in jest that extraterrestrials had held a picnic on this planet and had not cleaned up enough afterward. Life on earth was started by a bacterium surviving on a primordial cookie crumb.

A more serious variation on the theme of bacterial arrival by spaceship has been developed in some detail. Francis Crick, the co-discoverer of the double helical structure of DNA, and his longtime friend and colleague Leslie Orgel published a paper, "Directed Panspermia," in a space science journal in 1973. Crick later expanded this idea in a book, *Life Itself.* I fleshed out their theory by adding some specific details and presented it as Friday's Tale in the Prologue.

Francis Crick started with the observation that in a universe that is more than twice the age of the earth, "there is enough time for life to have evolved not just once, but two times in succession." He followed by noting that "although we cannot as yet give any powerful reasons why an origin elsewhere was much more plausible, it is rash to assume that conditions here were just as good as anywhere else." His theory was presented as speculation, an idea that had come in advance of any strong supporting evidence. In the book, he noted: "The kindest thing to state about Directed Panspermia, then, is to concede that it is indeed a valid scientific theory, but that as a theory it is premature."

One motive behind the publication of the book was to increase public awareness of the difficulties surrounding the origin-of-life question. Crick explained this to me during a private interview: "We thought of this theory but we're not all that sold on it. . . . The object [of the book] is to give the intelligent person an idea of what the *problem* is, and this is just a tag to sing it on. . . . Everybody, as they say in California, can *relate* to certain ideas and things like coming on an unmanned rocket—or even bacteria, they think they can relate to."

This relaxed, skeptical attitude stands in sharp contrast to the manner in which Sir Fred Hoyle and his own colleague of long standing, Professor Chandra Wickramasinghe, have presented their competing ideas. We encountered some of them as Saturday's Tale in the Prologue. These authors take their work quite seriously and present their ideas with conviction, even with certainty. They are astrophysicists, and start with a detailed knowledge of the stars and of the other component which makes up so much of the mass of our galaxy, the interstellar clouds. These latter objects, which we met briefly when we discussed the origin of our solar system, are less familiar to us, so we will turn our attention to them for a moment.

Stardust

When we think of outer space, we usually visualize multitudes of stars, perhaps with planetary systems, separated by absolute emptiness. In reality, such total voids do not occur. Isolated lonely atoms and molecules roam about in the space between the stars. Their average density is less than that present in the highest vacuums produced by laboratories on earth, but this density varies greatly. In some locations, the atoms or molecules are somewhat more clustered together, and interspersed with tiny solid particles of a size that would place them on the -7 level of COSMEL. These grains of dust, together with the atoms and molecules, compose the clouds. The density of matter in them is still quite low, but the clouds are so large, light-years in diameter, that one of them may contain 100,000 times the mass of our sun.

These clouds have been studied extensively by telescope. In some cases they appear as dark patches which obscure the light of the stars behind them. In other instances, the clouds can be observed directly, as some of them glow by their own light. They have been of particular interest to astronomers because new stars are formed within them.

A typical interstellar cloud can endure for millions of years.

The material within it may originate within existing stars and be released gently in a solar wind, or convulsively, in an explosion. Hydrogen and helium are the major components of clouds, as they are of the entire universe, but heavier elements are present as well. Carbon, oxygen, nitrogen, silicon, and other types of atoms are made by nuclear reactions within stars, and end up within the clouds. Ultimately these substances are concentrated in the planets.

Star formation takes place when local instabilities within a cloud cause sections of it to fall together, by gravity. The details of the process are not clear, and it is not understood whether planet formation is rare or common. To understand such processes fully, astronomers have been eager to learn the exact identities of the molecules and dust grains within the clouds. If a portion of an interstellar cloud could be sucked up in some cosmic vacuum cleaner and transported to a laboratory on the earth, this analysis would present little problem. This cannot be done, of course. The principal source of information concerning their chemical nature is the light and other forms of radiation that come from them or pass through them.

The light that we see with our eyes is only a small part of a much larger phenomenon called electromagnetic energy. This category includes such familiar forms of energy as X rays, ultraviolet and infrared light, and radio waves. These various energy forms are distinguished from one another by a property called wavelength, which can vary from thousands of meters for certain radio waves to less than one-trillionth part of a meter for cosmic waves. This variation in length is so great that it is best described using COSMEL. A wavelength of visible light would fall on the -7 level. Ultraviolet wavelengths are shorter but would occur typically on that level as well. A typical infrared wavelength might fall on the -6 or -5 level, and a microwave on the -3 level. Astronomers have analyzed the energy reaching us from the interstellar clouds at these various wavelengths, in an effort to identify the cloud components.

The best results have been achieved in the identification of small molecules, particularly with the aid of microwave spectra. The information received from the clouds consists of a series of many peaks, each representing a different wavelength. This series of numbers does not directly reveal what molecules are present, but it can be used in a process of inverse reasoning. An astronomer will guess that a certain molecule exists in the clouds, and then measure its microwave spectrum, or calculate it, if it is unstable under earthly conditions. If the peaks measured or calculated for the molecule are all represented in the spectrum from the clouds, then the conclusion is drawn that this substance is present out there. There is some opportunity for error due to coincidence, but this is small, particularly when a good number of peaks are associated with a molecule and all are present in the cloud spectrum. A 1982 account listed over fifty different molecules that had been detected in the clouds in this manner, enough to allow for some generalizations to be made. We must remember one important limitation, of course. A molecule must first be suspected, to be detected. There can be no total surprises discovered, due to the limitations of the procedure.

The molecules identified thus far all contain very few atoms; one has thirteen, another eleven, and the remainder nine or less. The elements present include hydrogen, carbon, nitrogen, oxygen, sulfur, and silicon. A variety of small organic molecules are present, some with unusual or incomplete bonds. Such substances would not endure on earth, but survive in the cold, uncrowded conditions of space. One familiar substance present in the clouds is ethyl alcohol. Its density in space is low, but the galaxy is so vast that the total quantity of alcohol present is enormous. We could prepare millions of martinis, each the size of the Pacific Ocean, using this alcohol.

Amino acids were expected in the clouds, and predicted in 1971, but none have turned up as yet, not even glycine, which has only ten atoms. Eventually, the simpler ones will be found, when more sensitive instruments are built, but they are clearly not abundant. The two largest molecules detected to date are rather strange substances, rich in carbon and depleted in hydrogen. They would not survive for long here. On the whole, the list of molecules found in space could best be described as a rather unearthly collection.

Despite this, some observers have used the list to support their own predestinist assumptions. The presence of simple organic compounds in the interstellar clouds is seen as an indicator of cosmic purpose, evidence that cosmic chemistry proceeds in the direction of our own specific biochemistry. In this respect, the clouds have served the function of a psychological Rorschach test, as each observer can see in them what he wishes. We will touch on this subject again in a later chapter. Now we must move on to consider the tiny dust particles in the clouds, which have proved to be an even more fertile stimulus for the imagination.

These grains of stardust are larger and more complicated than simple molecules, so less can be learned about them. Infrared and ultraviolet spectroscopy have been two important sources of information. We will linger with them for a time, as they are important to our study.

Infrared spectroscopy was a technique that I encountered often in my own graduate training in organic chemistry. At the time when I began my studies, it was the most important instrument used to explore the structure of moderately complicated organic products (now, of course, it has been overshadowed by more elaborate, and much more expensive, techniques). I learned that a pure substance would give a spectrum that resembled the skyline of a large American city: a sharp series of peaks and troughs. This spectrum, by itself, would not reveal the identity of the compound, but certain peaks in specific locations would give definite, but limited, information. Most of the spectrum, though, could not readily be interpreted, but served rather as a fingerprint, a clue to recognition. My graduate advisor, R. B. Woodward, was famous among chemists for his laboratory syntheses. He devised the first chemical preparation of many well-known substances, such as quinine, strychnine, and chlorophyll. Often, the infrared spectrum of the laboratory-made product was used as the definitive proof that the synthesis had succeeded. If it matched the spectrum of the natural material in every detail, peak for peak, inflection for inflection, then he would claim that the natural and prepared substances were identical.

To a chemist trained in this tradition, the sight of the spectrum obtained from an interstellar cloud is quite disappointing. It is smooth in most areas and relatively featureless, more like the outline of a ridge than a skyline. This shape reflects the technical difficulties involved in obtaining the spectrum, and also the fact that the dust grains are more likely to represent a diverse mixture rather than any single substance. From this very limited information, astronomers have tried to derive some inferences about the general nature of the grains, but have failed to agree even as to their organic or inorganic nature. Some workers have interpreted the spectrum in terms of a mixture of ice, silicates, and other minerals. Carl Sagan and his colleagues have suggested that they are tholins, a name they apply to brown, sticky, organic tars of the type made in Miller-Urey reactions. For an analogy to this situation, we must imagine an effort to identify an individual from a single photograph taken at a distance in a fog. The resolution does not even permit agreement on whether the person is a man or a woman.

Some additional information on stardust has been obtained from its spectrum in the ultraviolet energy range. I have become extremely familiar with this technique from my own research on nucleic acids. It is more limited than infrared spectroscopy. Almost every substance has an infrared spectrum, but only certain classes of compounds have ultraviolet spectra in the range that is usually used. Much less information is present in a typical ultraviolet spectrum than in an infrared one. Often only a single hump is seen. The data can serve for the identification of a substance only when the number of possibilities has been limited to very few, as established by other methods. The spectrum of the clouds contains only a single hump, in the most common location encountered in such spectra. Since the possibilities are literally almost anything in the universe, no firm conclusion can be made on this basis, although suggestions have been offered which range from a mixture of organic compounds to graphite (a layered form of the element carbon used in pencil "lead" and in some lubricants). The exact nature of stardust is an enigma, and may remain so until we can collect some of it and bring it home.

Most scientists would agree with this analysis, but Hoyle and Wickramasinghe have taken a different course and come to a series of definite but differing conclusions about the nature of the interstellar clouds. Their theories have involved the clouds in the origin of life. Before we consider them further, perhaps we should get to know these two gentlemen a bit better.

Two Dissenters

The more celebrated of the two, Sir Fred Hoyle, has had a distinguished career, and made many contributions to the field of astronomy. He and his colleagues first deduced the processes in which heavier elements are made from lighter ones within stars. He also helped develop the steady-state theory of the universe. In this theory, the universe has remained in its present condition indefinitely. As it expands, new matter is created continuously, to maintain a constant density. This idea has fallen out of favor with most scientists and given way to the Big Bang theory, in which the universe was created suddenly at a definite time, perhaps 10 or 20 billion years ago.

Hoyle was born in 1915 and spent most of his career in various faculty positions at Cambridge University. This career was marked by a number of controversies concerning university politics and administrative matters. In the mid 1960s, he resigned from the mathematics faculty and threatened to emigrate to the United States. He remained at Cambridge, however, as he had been appointed head of the newly formed Institute of Theoretical Astronomy. He left this post and quit his Cambridge professorship in 1972, after additional political quarrels. Hoyle made the headlines in 1975 when he asserted that one of his former Cambridge colleagues had received the Nobel Prize by claiming credit for studies performed by his assistant. Both the prizewinner and his assistant denied the charge.

These controversies stand small when compared to the many honors received by Hoyle, which include a number of awards and medals. He has also been past president of the Royal Astronomical Society, vice-president of the Royal Society, and a foreign associate of the U.S. National Academy of Science. He was knighted in 1972.

His talents extend beyond research to literature. He has written texts on astronomy and popular books on nuclear power and global climate changes. In addition, he has written a number of science-fiction novels, some in collaboration with his son, Geoffrey. In 1969, he prepared the libretto for an opera, *The Alchemy of Love*, written by American composer Leo Smit. According to the *New York Times*, the opera was concerned with "politics, love, and the abuse of genius."

Hoyle's younger collaborator, N. C. Wickramasinghe, is a native of Sri Lanka (Ceylon). He worked at Cambridge for some years, during the time of Hoyle's tenure there. More recently, he has been head of the Department of Applied Mathematics and Astronomy at University College, Cardiff, Wales. His specialty, appropriately, is the interstellar clouds.

The speculative theories of the two astronomers (we will refer to them as H. and W. for the remainder of this chapter) were launched during the period 1977–1981. They were presented fully in popular books, and certain aspects were treated in detail in more than a dozen papers in scientific journals. For years before that period, H. and W. had contributed a number of more conventional papers on the interstellar clouds to various journals. Their ideas appeared to shift and develop almost continually from 1977 to 1981, but for simplicity, I will sort them out into two distinct groups. The views presented in *Lifecloud* (1978) will be called the early theory, while those in *Diseases from Space* (1978), *Evolution from Space* (1981), and later works will be termed the later theory. We will consider each separately.

The Early Theory

In this version, H. and W. argued that specific molecules important in our biochemistry were present in outer space: "A molecule of formic acid and a molecule of methanimine could react to give the simplest amino acid, glycine, and there is every reason to believe that this will happen very extensively. So a quite complex prebiotic chemistry seems to be taking place already at the stage of prestellar collapse of dense interstellar clouds."

Formic acid and methanimine (another small organic molecule) appear on the list of substances identified in space. Glycine, as we mentioned, has not yet been detected in any amount, nor was any evidence concerning its presence presented by the authors. They went on, instead, to make additional claims concerning the presence of other biochemicals. The dust grains were identified, with certainty, as cellulose (we will consider the basis for this statement a bit further on). H. and W. summarized: "With the formation of these materials the foundations of biochemistry would appear to have been laid."

The interstellar clouds, however, did not play the most important role in their theory. This honor was reserved for another class of heavenly body, the comets. Comets, of course, are obvious candidates for the lead in any drama of outer space, because of their spectacular appearance. These objects, with their shining heads and long tails, have turned up in the night sky at various times in human history and inevitably made a profound impression. The sight of a comet was taken as a signal that a very important event was about to occur. "When beggars die there are no comets seen; The heavens themselves blaze forth the death of princes," wrote Shakespeare in *Julius Caesar*.

We appreciate now that the movements and history of these visitors far transcend our current earthly concerns. Their heads are small, perhaps ten kilometers (six miles) across, but their far less substantial tails may extend for millions of kilometers. Great multitudes of them follow distant orbits beyond Saturn, and have done so since their formation at the time when our solar system took its present form. Occasionally, one or another of them is perturbed by an event that moves it into a new orbit, one that periodically brings it much closer to our sun. Comets are made largely of ice, and other substances that convert readily to gaseous form. As they approach the sun, these materials evaporate, forming the tail. In 1985 and 1986, Halley's comet has once again provided us with the most famous display of this type.

Astronomers have been eager to learn what other substances may be present in comets, apart from ice, but have had only limited opportunities to observe their spectra. They have identified a few of the simpler molecules seen in the interstellar clouds, but no species containing more than six atoms has been reported.

In their *Lifecloud* scenario, however, H. and W. claim that at the time of their formation comets absorbed large quantities of biological material from the clouds. The authors reproduce in that book the list of molecules identified in comets, but add to it the words "polysaccharides and related organic polymers." They support this only with the statement: "A better explanation in our view is that many of the radicals which are observed are dissociation products of organic polymers, such as the polysaccharides." (The word "radicals" refers to unstable organic molecules with incomplete bonds.)

The theory continues, describing the state of the early earth to be somewhat like the present moon, with no atmosphere. This deficiency was remedied when comets made numerous soft landings, to bring in the necessary supplies. H. and W. mention the alternative hypothesis, accepted by most geologists, that our atmosphere was formed from gases released from within the earth. They dismiss it with the words: "This explanation, however, is open to serious objections. To begin with, it is not based on any evidence."

Once the atmosphere was in place, the comets could ferry in the additional ingredients needed to stock the prebiotic soup. We will cite the authors again: "If interstellar space is full of prebiotic molecules . . . it is almost self-evident that the origin of life on Earth merely involved a piecing together of interstellar prebiotics." However compelling this conclusion may have been, it was set aside a few pages later in favor of another alternative: a better place than earth for the cooking of the soup was the interior of the comets themselves. This happened repeatedly, within many comets, to produce bacteria and viruses. "Then about four billion years ago life also arrived [on earth] from a life-bearing comet." Such deliveries have continued since that time, stimulating evolution.

The Response of the Scientists

These provocative ideas, put forward by a famous astronomer and his colleague, could not pass unnoticed. The most negative of many criticisms was perhaps that of biologist Lynn Margulis. She stated: "The book is flamboyantly irresponsible. Its theme moreover is entirely contrary to the considered opinion of most workers in the field, if 'origins of life' can be considered a field. The book is wanton, amusing, promiscuous fiction." She drew

the conclusion that Hoyle had used his name and position to secure publication of his ideas as a book, rather than face scientific
critics. In this, at least, she was incorrect, for much of his theory
had been presented in scientific journals.

The professional response to this theory was by no means
wholly negative, however. Astrophysicist John Gribbin wrote in
his own book *Genesis* that their hypothesis "provides the most
complete explanation of what is going on in the dust clouds of
space." Whatever the difficulties, "something along these lines
will eventually become the established view." Hoimar von Dit-
furth, the author of a best-selling book in Germany on evolution
and the origin of life, expressed similar sentiments, adding: "Far
more complicated molecules up to complete amino acids and ri-
bonucleic acid itself spontaneously come into being in cosmic
gas clouds." Astronomer W. M. Irvine and his colleagues at the
University of Massachusetts wrote in a 1980 *Nature* paper: "It
follows . . . that comets may contain quite complex organic molecules, and that comets may have played a role in the origin and
conceivably even in the subsequent evolution of terrestrial life."
Their paper, however, dealt only with possible temperatures
within comets, not their contents. They relied upon the H. and
W. theory for the biological connection.

The early theory of these authors had started a small band-
wagon among like-minded scientists and writers. There would
have been far fewer adherents had the theory been presented
only in popular books, but this was not the case, as we have seen.
Many technical papers appeared in refereed professional journals, and we must turn to these papers to examine the strength of
their claims.

The H. and W. Papers

Hoyle and Wickramasinghe have argued their theories using
data from various sources. To consider all of their presentations
in detail, we would need an entire book, rather than a chapter.
We will sample their work instead, focusing our attention mostly
on a single important claim, and allowing that one to represent
their general approach.

The claim made by H. and W. that appears to have made the

greatest impression was the identification of the interstellar dust grains as cellulose, on the basis of their infrared spectra. This polysaccharide, perhaps the most abundant biological product on earth, is the principal material used in the construction of wood and other plant substances. We encounter it in cotton and paper; it is before our eyes right now in the pages of this book.

This identification had such impact because cellulose is an amazingly specific material, produced on earth only by biological processes. Any purely chemical path capable of yielding this substance would be expected to yield an immense number of other products as well, affording a complex mixture.

Let us consider the steps that would be needed to prepare cellulose. The small molecules in the clouds could react with one another in many ways, producing many classes of organic compounds. From our knowledge of organic chemistry, we would not expect sugars to be abundant; even if they were, many hundreds of sugars could be formed. Glucose would of course be among them, but to get to cellulose, glucoses would have to seek each other out exclusively, ignoring the many other sugars as well as the more numerous molecules that are not sugars. Even if this unlikely event should take place, further complications would still lie ahead. A carbohydrate chemist has calculated that there are 176 different ways, all chemically reasonable, that three glucose units can combine together. To progress toward cellulose, only one of these alternatives must be selected. Further, with each additional glucose that was added, the proper choice must be made from 20 new possibilities. This selectivity would be needed until dozens of glucoses had been strung together, producing a small cellulose unit.

If H. and W. had suggested that the dust grains were made of a complex organic mixture, or a nonspecific irregular material such as tholin, they would have attracted little attention. The cellulose claim, however, appeared almost miraculous. To accept it, we would have to assume either that some predestined force was directing interstellar chemistry along paths that led to our own biochemistry or that biological creatures related to us were already loose in the cosmos. A claim of this sort, for credibility, would need overwhelming documentation, with details sufficient to distinguish cellulose from the literally astronomical number of other possibilities. Any well-equipped laboratory on earth could quickly establish this identification or refute it, if given a sufficient supply of stardust. H. and W. proposed to do it, however, on the basis of the largely featureless infrared spectrum that we have described.

We can trace the evolution of their thoughts by following their papers in chronological order. In one published in *Nature* in 1969, they claimed: "Interstellar grains may be a mixture of graphite particles formed in carbon stars and of silicates in oxygen-rich giants." By 1974, Wickramasinghe had changed his mind, and become attracted to the virtues of POM (polyoxymethylene), an organic polymer (a product made of subunits) unrelated to biology. He wrote: "POM grains are thus able to explain all available interstellar extinctions. In view of the spectral identifications presented here, POM grains must clearly be regarded as a strong candidate for the main component of interstellar dust."

This candidacy had been swept aside by early 1977. H. and W. averaged the spectra of eighteen different organic polymers of various kinds and found that the composite provided the most satisfactory fit to the infrared spectrum of an interstellar cloud. The reign of their new champion was incredibly brief, however. In the very last paragraph of that paper (which may have been added after their initial submission), they noted that this organic mix was less significant in the cosmos than their new favorite, polysaccharides. This comment was supported by a reference to a new paper, one which had not yet even been submitted for publication. Obviously, their thoughts were developing rapidly at this time.

Lifecloud affords some insight into this process: "Early in 1977 we ourselves became convinced that it would be far better if a single chemical substance could be found to explain all the main features of infrared radiation from astronomical sources." They turned to biological material, including a waxy component of spores and pollen, but could not find the match they wanted. Then suddenly, a new thought occurred to them:

It was only then, somewhat belatedly, that we asked ourselves a crucial question: What are the infrared properties of the *most abundant* terrestrial organic substances, cellulose? A dash to the library, and we found to our amazement that laboratory measurements for cellulose over the wavelength range from 2 to 30 microns showed just the absorption bands we were seeking. Moreover, cellulose was free of unwanted bands. This close agreement . . . convinced us there was a strong *prima facie* case for saying that interstellar dust consists in the main of cellulose, or of some related polysaccharide.

No spectrum measured under laboratory conditions, of course, would be expected to match exactly the one from the clouds. Earthly infrared spectra are full of fine detail—bumps, dips, and shoulders—that are absent in the interstellar one. It is just this rich detail that makes them valuable for identification purposes. H. and W. adjusted the laboratory spectrum of cellulose, using a method they devised themselves, to compensate for the differences between the conditions present on earth and those in outer space. After this adjustment, a better match was obtained.

To appreciate the significance of this procedure, we must return to the analogy of the distant figure photographed in a fog. Imagine that an observer claimed that the figure was definitely Ronald Reagan. He produced a photograph of the President and then painted out the features to mimic the effects of the fog. This adjusted version was then compared to the fog photo, with comments about their similarity. For example, both figures clearly had two arms and two legs. If we were impressed by this comparison, then we would be ready to accept the argument of H. and W.

The authors themselves were not quite happy with their claim that the interstellar grains were made of one material, cellulose. They wrote in another paper: "We might expect an abiogenic synthesis . . . to lead to the formation of a hybrid mix of stable polysaccharides rather than a single polysaccharide." A new comparison was made, using an adjusted, smoothed-out average spectrum of four polysaccharides selected by them. They acknowledged that many other combinations could be concocted that would also match the cloud spectrum, but added: "A hybrid mix of organic solids which meets this requirement cannot, of course, be excluded. But such a mixture will, of necessity, be contrived and *ad hoc*."

Soon, however, they had to resort to an *ad hoc* contrivance of their own. Several months later they noticed that their close fit still contained "two significant points of departure." One departure could be cured, however, if they presumed that a certain hydrocarbon, selected apparently on its ability to improve the fit, was also present in the clouds. They did not consider that this procedure weakened their overall argument, but rather that it "points strongly to the identification of hydrocarbons of this type which may be associated with polysaccharide grains in interstellar space."

As we might expect, this series of presentations by H. and W. provoked a wave of detailed technical rebuttals. The scientists involved in this enterprise could have saved their energy. Within a short time, the authors moved to a very different conclusion concerning the grains: they were not a polysaccharide mix, with or without hydrocarbons, but rather freeze-dried bacteria and algae.

The arguments supporting this new position were published in an astronomy journal. *Nature*, which had patiently printed all of their previous revelations concerning the dust grains, may finally have lost patience at this point. Infrared spectroscopy played a diminished role in the new wave of arguments, but H. and W. attempted to maintain some link to the past: "We assume the optical properties of the biological component to be satisfactorily represented by our laboratory data for cellulose."

Bacteria, unlike trees, are not made of cellulose, however. Their outer cell walls *do* contain a polysaccharide component, but a very different one. These walls also contain amino acids and other important substances. If these additional components made no difference in the spectra, then all of the identifications that H. and W. had made using infrared spectra were valueless.

Much greater evasive maneuvers were needed for them to reconcile their bacterial claim with the ultraviolet spectrum of the clouds. The principal ultraviolet-absorbing substances in bacteria are proteins and nucleic acids, and their spectra differ greatly from that of the clouds. H. and W. attacked this problem with the following statement: "Unfortunately we do not have the ultraviolet spectra for intact biological systems, so we are forced to approach this question in an indirect manner." They consulted a 1964 book by A. I. Scott, *Interpretation of the Ultraviolet Spectra of Natural Products,* and selected nine tables from it. They claimed that the spectra of the 186 molecules listed in these tables, when averaged together, produced a composite spectrum close to that measured from the clouds. In their words: "This close agreement between our calculated mean absorption curve and astronomical data ... gives strong support to our contention that chromophores (absorbers) in bio-molecules dominate the interstellar absorption at these wavelengths."

Fortunately, I was able to find a copy of Scott's book, covered with dust, on a bookshelf in my office. After inspecting the tables H. and W. had cited, I could only conclude that they had not troubled to read the entries when they computed their average. If they had done so, they would have noted the presence of many compounds irrelevant to bacteria and even to biology, which were included to illustrate theoretical points. The table of pyrimidines that they used (pyrimidines are a subclass of base), for example, contained fifteen items, but omitted one of the two normally present in DNA. The majority on this list had no biological significance at all.

As I completed this inspection, I noted one final irony. H. and W. had not even totaled their compounds correctly. Only 153 entries were listed in the tables cited by them, not 186.

I have spent this time with the spectra to give the reader some idea of the quality of the science contained in these later papers. In doing so, however, I have taken us beyond the limits of the early theory, which did not propose that the dust grains were bacteria. We shall move on, to consider their subsequent thoughts more fully.

The Later Theory

In a series of popular books published from 1979 to 1981, Hoyle and Wickramasinghe developed a second theory, which differed in many respects from the earlier one. We have summarized the highlights in Saturday's Tale in the Prologue. The reversal in attitudes between the theories is remarkable, considering the short time that separated them. For example, the earlier version had accepted the origin of life within a prebiotic soup, stating: "The principle of this process is not in question" and "it is now virtually certain, moreover, that similar experiments in biological assembly occurred on innumerable occasions in many other places in the universe." This certainty notwithstanding, Hoyle wrote three years later: "Another fuddled notion is that life began here on Earth in a thin brew of organic material. The mystery is why grown men and women have allowed themselves to be persuaded into such beliefs, in spite of there being a considerable body of fact running against them." The soup was set aside in favor of a Creator.

In *Lifecloud*, H. and W. spoke of Darwin's theory in the following way: "When it was published in 1859, Darwin's *On the Origin of Species* met emotional opposition from almost every quarter. ... Darwin's theory, which is now accepted without dissent, is the cornerstone of modern biology. Our own links with the simplest forms of microbial life are well nigh proven."

A bit later, however, they wrote the following: "These conclusions dispose of Darwinism, which cannot produce genetic changes quickly.... The speculations of *The Origin of Species* turned out to be wrong as we have seen in this chapter.... Nobody seems prepared to blow the whistle on Darwinian evolution. If Darwinism were not considered socially desirable and even essential to the peace of mind of the body politic, it would of course be otherwise."

In addition to the repudiation of many of their previously held convictions, the authors brought many new ideas into their later work. In extended arguments, they traced the influence of disease, brought in from outer space in various ways, on the course of biological evolution and human history. These topics fall outside the major theme of this book, yet I cannot resist the chance to present a few selections.

Cancer, for example, can result when a set of genetic instructions from space, intended to promote the budding of yeast, is received accidentally by animal or plant cells. "The phenomenon of cancer is an inevitable consequence of the present ideas."

Many developments in history were also caused by diseases

from space. "The explanation of why classical armies were superior to medieval ones lies of course in disease, with which the medieval age was riddled. . . . We also attribute the rise of Christianity to the same disease-filled epoch."

As we have seen, H. and W. used spectroscopic data to fortify their claim that space offers a rich harvest of biological organisms. They pointed out, however, that an alternative proof can be found if we simply consider our own noses. A few million years ago, when our apelike ancestors lived in the forest, their noses were little more than two holes in their faces. Then they moved to open ground, which was hazardous, "unlike dense unbroken forest, which afforded effective protection against the rain of pathogens from the sky." Strong selection pressure resulted for the development of a nose, as a protection against disease caused by sniffing in dangerous raindrops.

Returning from these lesser themes to the origin of life, we mentioned above that H. and W. had set aside any consideration of spontaneous chemical paths in favor of a Creator for our kind of life. Their selection, however, was not any entity mentioned in a conventional religion but one described only by themselves. They wrote: "While many are willing, and some are anxious, to postulate an ultimately surpassing intellect, God, few are happy with the thought of intelligences intervening at levels between ourselves and God. Yet surely there must be such intelligences. It would be ridiculous to suppose otherwise."

Our own immediate progenitor was "an extremely complex silicon chip." Such chips, so vital in modern computers, had the computational power needed to design the first bacteria. This was done not for altruistic purposes, but rather with the intent that the bacteria evolve into beings capable of constructing computers, thereby spreading silicon-chip life throughout the universe.

As we have seen, the earlier theory of H. and W. was accompanied by technical papers, which attempted to demonstrate, however poorly, the scientific basis for their claims. As their suggestions became more and more extravagant, however, so did the quantity of technical supporting data diminish. The most grandiose statements were put forward virtually on their own authority. Little patience was shown for readers unwilling to accept them on this basis. The authors wrote at one point: "Doubtless there will be persons who never take a positive statement like this on trust, persons who would continue to argue, even as the snow closed over their heads, that an avalanche was not really bearing down upon them."

But if data are not involved, how then do H. and W. arrive at their conclusions concerning, for example, the hierarchy of intelligences that govern the universe? In finishing their chapter on this topic, they state: "The connections of the sequence are more likely to be restricted to those sudden flashes of perception that have made so much difference to all the main trends of human thought, the conversion of Paul on the road to Damascus."

With this endorsement of revelation as the source of knowledge, the authors completed their own transformation, from science to religion. By stages they moved from their technical papers of the 1960s and early 1970s, in which plausible, sober, and possibly correct inferences were made concerning the most likely chemical composition of the interstellar dust particles, to their essentially religious position of the 1980s.

In this final position, conclusions were drawn concerning the nature of the dust and the entire universe which derive from their own inner convictions, rather than any impartial considerations of experimental results. Only those arguments and evidence were produced that would support their preestablished position. When we see a remarkable transformation of this type, particularly in a scientist as distinguished as Sir Fred Hoyle, we cannot help wondering at the circumstances that provoked it. He has not shared his inner thoughts with us, but has left some clues in his writing.

Hoyle's biological and theological convictions are not new ones, but are connected in a unified vision which includes the steady-state theory of the universe. In this theory, the universe had existed for an indefinite, very long period of time. This time was needed to allow for the development of the incredible information content present in our type of life, and the even greater amount in the more intelligent beings above us in the cosmic scheme. The oldest, most intelligent entity is the universe itself.

"The steady-state conception is of a universe that contains within itself its own perception, its own divinity, as one might say." This conception, however, challenges the basic tenets of both conventional science and Judeo-Christian religion, which agree that the universe was created suddenly, from nothing, at a definite time. In Hoyle's view, his theory was attacked by astronomers "with an almost insensate fury" because their basic belief system had been threatened. "I used to remark that the community of astronomers lived in perpetual terror that one day it might stumble inadvertently on something important, a remark that did not greatly enhance my popularity," wrote Hoyle.

Hoyle has suggested that his biological ideas were formed late in his career, after the steady-state theory. Some of my colleagues have suggested privately that he simply took leave of his senses at that time. However, another source of information suggests that this is not so, and that he had incubated his entire belief system for a longer period of time. We need only consult a science-fiction novel written by Hoyle that was published in 1957.

The Black Cloud

In the novel that bears this title, a dense, compact interstellar cloud enters our solar system and engulfs the earth. Direct sunlight is cut off on our planet, provoking an abrupt temperature drop and worldwide catastrophe. A group of scientists meets in Britain to consider the emergency, and one of them deduces that the cloud is alive, stating: "I expect the chemistry of the cloud to be extremely complicated—complicated molecules, complicated structures built out of molecules, complicated nervous activity."

The cloud not only is alive, but far surpasses humans in intelligence. It communicates with the scientists, stating its surprise at finding intelligent life on a planet. Space is a far superior place for the assembly of biochemicals.

The cloud is willing to share its theological opinions, as well as its scientific expertise, with the humans: "By and large, conventional religion, as many humans accept it, is illogical in its attempt to conceive of entities lying outside the universe. Since the universe comprises everything, it is evident that nothing can be outside it." The cloud senses the existence of higher intelligences within the universe, and in the end, sets off in search of them.

Thus this early novel contains the heart of the philosophy behind Hoyle's final position, though it predated the evidence cited in support of it. Hoyle toward the end of his career had become willing to state as fact convictions that he had earlier presented in his fiction. The scientific achievements and the eccentric belief system were different facets of the same individual.

This situation is not unique in science. Early in 1983, *The New York Times* carried a story under the headline "What Happens When Heros of Science Go Astray?" The article, citing the research of historian Frank E. Manuel, discussed Isaac Newton and naturalist Alfred Russel Wallace, among others. (Ironically, Hoyle had mentioned both with approval in an article published just before that time.) Newton had conducted an alchemical quest for mysterious elixers and occult powers, while Wallace had been involved with séances and other attempts to communicate with the dead.

A psychologist, Ray Hyman, had also examined these cases. At first, he thought that Newton and Wallace had undergone pathological changes; they had simply gone mad. After much study, however, he concluded that their reasoning had stayed the same. The same personality traits that had led to success also caused failure. Perhaps the case of Sir Fred Hoyle can be added to this list.

Our diversion into outer space has not advanced us on our search for the origin of life, but illustrates the difficulties that come up once the skeptical approach necessary to science is abandoned. Hoyle and his colleague had started with the consideration of experimental data, but ended with a mythology of their own, which they chose to call science. It is fitting that in this final position they were embraced by another group that had reached it from another direction. This group, the Creationists, started with Scripture, then sought experimental evidence to support their predetermined position. In applying the term "science" to their belief system, they perhaps reached the ultimate limit in confusing the two disciplines. We shall consider their case next.

ten

Creationism: Religion as Science

"Public schools within this state shall give balanced treatment to creation science and to evolution science." So began a proposal introduced in the Arkansas legislature in February 1981, fittingly called "Balanced Treatment for Creation Science and Evolution Science Act." These terms required definition, of course. A later section of the act stipulated that " 'Creation science' means the scientific evidences for creation and inferences from those scientific evidences," while " 'evolution science' means the scientific evidences for evolution and inferences from those scientific evidences."

To give substance to these positions, a series of six statements was made for each doctrine, setting forth the opposing viewpoints. Creation science, for example, supported the "sudden creation of the universe, energy, and life from nothing," while evolution science favored the "emergence by naturalistic processes of the universe from disordered matter and emergence of life from nonlife." Creation science endorsed "a relatively recent inception of the earth and living kinds," while evolution science favored "an inception several billion years ago of the earth and somewhat later of life." Three other points dealt with aspects of Darwinian evolution, and another one with the occurrence or nonoccurrence of a worldwide flood.

This bill, Arkansas Act 590, encountered little opposition in the legislature, and was passed by large margins in both houses the next month. Two days later the governor, Frank White, signed it into law "with much ostentatious flourish, without reading it and against the advice of a legislative assistant who had," according to an account in *Science*.

Should we have objections to this act at this point, before we have considered the evidence on any of its six arguments? The answer must be yes, as the very construction of the opposing viewpoints raises problems. Would students have the freedom of diners ordering à la carte, to choose freely among items on the menu? In the Gallup poll mentioned in the Prologue, 38 percent of the respondents endorsed the statement "Man has developed over millions of years from less advanced forms of life, but God guided this process, including man's creation." This substantial group holds certain beliefs listed under evolution and others linked with creation. Is the viewpoint of this group to have no identity of its own?

Other schools of thought fare even more poorly, as they cannot be reconstructed by any combination of ideas listed under the two defined positions. A theory which we can call "Hoyle science," for example, supports an indefinitely old universe; the gradual development of our life, not by evolution, but by genetic messages sent in comets; and a hierarchy of creators, each creating the one below it in the sequence. Hoyle has had a notable and honored scientific career, and aspects of his theory have been published in respectable scientific journals. Should not these ideas be represented in the Arkansas act, if we are to have a "balanced" approach to creation and evolution?

Without even exploring the content of the issue, we can see that those who wrote the act have stacked the deck before dealing the hand. They have selected six points out of a much larger number that arise in the areas of origins and evolution, and grouped them together in a way that summarizes their own philosophy. No logical or scientific connection between the points exists, except that they have coexisted historically in the belief system of the Creationists. There would be no other reason why a belief in a recent worldwide flood should be connected to a belief in the sudden creation of the universe from nothing.

Having thus given special status to one belief system, the act's authors then combined everyone else into a single additional system, "evolution science," which had the opposite point of view on all six questions they raised.

It was not surprising that Arkansas Act 590 was almost immediately challenged in federal court in a suit, *McLean* v. *Arkansas Board of Education,* which was tried before Judge William R. Overton. The challenge was made by a group of twenty-three organizations, including the National Association of Biology Teachers, the United Methodist, Presbyterian, Roman Catholic, Episcopal, and other churches, the American Jewish Committee and other Jewish organizations, and the American Civil Liberties Union.

Before we proceed to the scientific issues and the resolution of the case, it is worth considering the historical background of this unusual conflict between a state government and a striking alliance of virtually the entire scientific and religious communities of the United States.

The Creation of "Creation Science"

Mythological and scientific approaches to reality have both been used by mankind since earliest times, and not infrequently, differing views have arisen from the two systems. The modern controversy that led to the above case came from a unique event, however—the publication by Charles Darwin of *The Origin of Species* in 1859. Darwin's concept that man was not created by God directly but evolved from lower organisms undermined ethical systems built upon man's special and direct connection to God.

The Origin of Species did not deny religion directly, of course, but only certain literal accounts given in the Bible. Darwin himself had said: "It seems absurd to doubt that a man may be an ardent theist and an evolutionist." Most religions of the Judeo-Christian tradition came to terms with evolution by regarding some parts of the Bible as an allegory, an account in which spiritual meaning is conveyed in terms of symbols not intended to be taken as literally true. For example, each "day" of the seven days of creation might be considered as a much longer period, involving many millions of years.

One branch of Christian belief in the United States, evangelical Protestant fundamentalism, took a different turn. This group believed that the Bible was inerrant, literally true as written. Darwin's theory was therefore incorrect, and the evidence supporting it flawed and in error. Further, the spread of this incorrect view was eroding the ethical basis of religion and promoting the destruction of our civilization. A statement from a recent Creationist work by Henry Morris and Martin Clark, *The Bible Has the Answer*, summarizes this view: "Evolution is not only anti-Biblical and anti-Christian, but it is utterly unscientific and impossible as well. But it has served effectively as the pseudo-scientific basis of atheism, agnosticism, socialism, fascism, and numerous other false and dangerous philosophies over the past century."

The movement gained considerably in vitality and influence in the period following World War I. The enormous loss of life and destruction of property in that war testified to the decline of morality in modern times and shattered illusions about the future of Christian society. A counterattack against evolution appeared essential to the biblical literalists, and they gained an important asset for this purpose. William Jennings Bryan, the noted politician and orator, and three times unsuccessful presidential candidate, actively joined their cause.

Bryan had been influenced by books about World War I that claimed a connection between Darwinism and German militarism. He became critical that the teaching of evolution in schools was undermining the religious beliefs and morality of young people. He joined efforts, successful in Tennessee, Arkansas, and three other states, to ban the teaching of evolution, stating: "The movement will sweep the country and we will drive Darwinism from our schools."

The Tennessee law was tested when a high school teacher, John Thomas Scopes, was tried for teaching human evolution in 1925. Bryan joined the prosecution in that case, while a celebrated lawyer and agnostic, Clarence Darrow, served with the defense. The event, held in Dayton, Tennessee, drew worldwide attention as the "Monkey Trial." Scopes was found guilty and fined $100, but the conviction was later overturned. The technical outcome was overshadowed by a dramatic confrontation in which Bryan took the witness stand and was cross-examined by Darrow on his religious and scientific beliefs. Bryan, quite flustered, admitted that he himself deviated from a completely literal interpretation of the Bible. The days of creation may well have lasted more than twenty-four hours each, he conceded, and the earth might well be more than a few thousand years old. Exhausted by heat and the stress of the trial, Bryan died within a week of its conclusion.

The trial had provoked great support for the evolutionists, in terms of favorable publicity, but the fundamentalists had achieved many of their objectives. To avoid controversy, many biology texts used in high school were greatly contracted in their coverage of evolution. Scopes, for example, had taught from the 1914 edition of Hunter's *A Civic Biology*, which contained three pages on evolution, with supportive material elsewhere. The 1926 version of this text removed most of the discussion of evolution; the word itself was dropped from the index.

The anti-evolution laws lingered on the books for years. The Arkansas version was declared unconstitutional only in 1968. The fundamentalist movement itself, which came to be called Creationism, dropped from public view, however. Divided into factions, it declined during the period 1930–1960. As it did so, the teaching of modern evolutionary theory gradually resumed. The launch of the spacecraft Sputnik by the USSR in 1957 provoked a wave of self-doubt about the adequacy of science teaching in this country. The National Science Foundation funded programs to improve curricula and textbooks. Evolution reappeared as a major theme in high school biology.

Once again Creationism revived to do combat. This time the unifying figure was a little-known civil engineer, Henry M. Morris, who had come to the conclusion that "God doesn't lie." His

1961 book, *The Genesis Flood* (with J. C. Whitcomb, Jr.), reaffirmed the literal interpretation of the Bible. He argued that "the real issue is not the correctness of the interpretation of various details of the geological data, but simply what God has revealed in His Word concerning these matters." A new feature had been introduced, however. Footnotes were included, and the work resembled a scientific publication in format. Thus, scientific creationism was born.

Literal interpretation of the Bible gained new energy, and in 1963 the Creation Research Society was founded. To attain regular membership in the society, applicants were required to have an advanced degree in a field of science and to sign a pledge which included the following statements:

1. The Bible is the written Word of God, and because we believe it to be inspired throughout, all of its assertions are historically and scientifically true in all the original autographs. To the students of nature, this means that the account of origins in Genesis is a factual presentation of simple historical truths.

2. All basic types of living things, including man, were made by direct creative acts of God during Creation Week as described in Genesis. Whatever biological changes have occurred since Creation have accomplished only changes within the original created kinds.

Further statements in the pledge confirmed the worldwide flood, Adam and Eve, and the divinity of Jesus Christ. In essence the organization attracted scientists willing to forsake the practice of their profession in certain areas, and accept instead explanations based on the word of authority alone. By 1981 the society had 650 regular members (those with advanced degrees).

For the general public, the Institute for Creation Research (ICR) and the similar Creation-Science Research Center (CSRC) were founded, both in San Diego, California. The former is the more prominent one today, with Henry Morris as director and Duane Gish, a biochemist with a Ph.D. degree from Berkeley, as vice-director. The opinions of these newer groups reflect those of earlier times. The CSRC maintains, for example, that evolution promotes "the moral decay of spiritual values which contributes to the destruction of mental health and . . . [the prevalence of] divorce, abortion, and rampant venereal disease."

New marketing approaches would be needed for the old package, however. By the 1970s, the fight to outlaw the teaching of evolution had been lost in the courts. The Creationists decided to settle for the next-best alternative: to have their doctrines taught in the schools, alongside evolution.

One obstacle existed to this approach. The U.S. Constitution forbade the teaching of religion in the public schools. This country was founded on a principle of neutrality between competing religions. Throughout history, force had often been used by one religious group to impose its opinions on others. To avoid this possibility, the authors of the Bill of Rights elected to leave the public arena vacant. So if the Creationists wished to present their doctrines in the public schools, some disguise would have to be provided. To achieve this objective, they drew up new versions of their texts in which references to God and other obviously religious aspects were deleted, and the phrase "creation science" was introduced.

One important organizer of the new legislative offensive was Paul Ellwanger of South Carolina. He outlined the new strategy in a letter to an associate, which turned up at the Arkansas trial: ". . . we'd like to suggest that you and your co-workers be very cautious about mixing creation-science with creation-religion. . . . Please urge your co-workers not to allow themselves to get sucked into the 'religion' trap of mixing the two together, for such mixing does incalculable harm to the legislative thrust."

In urging their approach upon legislators, Creationists appealed to concepts derived from the American legal system. Students should be exposed to both sides of a question, as in a trial. Freedom of speech, academic freedom, and just plain fair play required that their side be heard. Science presented a forum in which all kinds of data could be presented and every opinion expressed. The *Bible-Science Newsletter* advised the following strategy:

Sell more SCIENCE . . . who can object to teaching more science? What is controversial about that? . . . Do not use the word 'creation.' Speak only of science. Explain that withholding information contradicting evolution amounts to "censorship" and smacks of getting into the province of religious dogma . . . you are for science; anyone else who wants to censor scientific data is an old fogey and too doctrinaire to consider.

The freshness of this new approach carried the day, for a time. School boards were impressed. In the 1980 presidential campaign, candidate Ronald Reagan stated: "Whenever Darwinism is presented in the public schools . . . the biblical story of creation should also be taught." Early in 1981 the legislature of Arkansas, after a quick flood of lobbying by the Moral Majority, evangelists, and other groups, passed the bill. An article in *Science* described the governor's attitude in the following way: "White, who describes himself as a 'born-again' Christian, owed political debts to the Moral Majority for their efforts in helping him get elected, and he saw his endorsement as a way of paying some of these." Whatever his motive, the bill became law.

Debates, usually held on a university campus, have been a very effective tool for advancing the Creationist viewpoint. More than a hundred encounters have been held over the last decade or so, with Henry Morris and Duane Gish quite often representing the Creationist side and professors from the locality serving as the advocates of conventional science. Audiences have numbered up to five thousand, and the Creationists have held their ground quite well. Gish in particular has performed impressively. One admiring colleague commented that he "hits the floor running," just like a bulldog. Gish himself added, "I go for the jugular vein." An obviously partisan account of the debates, published by a Creationist press, has the title *From Fish to Gish*. A series of panels on the front cover show a fish evolving to a lizard, a woodchuck, an ape, a caveman, and finally into Gish, who catches the fish!

Gish and Morris were not the first Creationists to "haul in" their professional opponents. Harry Rimmer (1890–1952), a Presbyterian minister, self-styled "research scientist," and biblical literalist, had worked the same circuit a half-century earlier. He gave many lectures, and by his own estimate, never lost a public debate. After one contest with an evolutionist, he wrote home: "The debate was a simple walkover, a massacre—murder pure and simple. The eminent professor was simply scared stiff to advance any of the common arguments of the evolutionists, and he fizzled like a wet fire-cracker."

Many reasons can be given for this good performance of Creationists, then and now. The very format of the debates gives them everything they sought in the Arkansas law. The issues selected publicize their own point of view, and give it equal status with all opposing ones combined. The debate format itself distorts the practice of science, which is not constructed on an adversary basis. Science is defined by its method, not by any position to be defended. To join the word "science" to any fixed dogma, as in evolution science or creation science, is a self-contradiction. Only one science exists, and it has no built-in point of view. The weight of evidence determines the conclusions, whatever they may be. In a debate, the practice of equal division of time between two sides may disguise an enormous disparity in the weight of evidence supporting opposing explanations.

Factors unrelated to the merits of the positions have also played a role in the debates. Creationists come as the underdog, with a novel point of view, facing the dull, conventional establishment. Their speakers have practiced in many previous debates, anticipate the questions that come up, and are at ease. The scientists are specialized, with extensive knowledge of the technical details of a limited area, but often with little grasp of wider issues or the philosophy of science. They have not learned how to handle themselves in debates. In one televised encounter, Gish opposed Russell Doolittle, a biochemist from the University of California at San Diego. Doolittle lost control of his time, grew flustered, and lost the debate, according to the media. He later commented that he had made a fool of himself. Whether he had or hadn't, his performance reflected his debating skills rather than the merits of the issue.

The Skeptic wishes to interrupt the narrative at this point. He would like to know more of the content of Creationist positions. What material do they use in their texts, newsletters, debates, and the quarterly journal that they publish, other than quotations from the Bible? They cannot further describe the activities of the Creator, or document his properties with suitable experiments. What, then, fills their literature?

Supporters of mythology search for evidence that supports their position, while taking care not to imply negation if their search should be unsuccessful. Creationists sponsor expeditions to Mount Ararat to look for Noah's Ark, for example. Such activities represent only a minor part of their efforts, however. Their major enterprise is the criticism of conventional science, in areas where it threatens their doctrines. Creationists collect anomalous results, and criticize faulty procedures and logic used by scientists. When this is done responsibly, such criticism actually serves a useful purpose, in helping to detect errors in the scientific literature. The Creationists err, however, in presuming that this activity supports their own position.

Anomalies and artifacts exist in every scientific field. A certain level is expected, as part of the normal practice of science. Their existence cannot support the Creationists' main idea, which lies outside of science, invulnerable to negation, but also incapable of affirmation, by scientific experiments.

As critics of conventional science, with no body of experimental work of their own to defend, the Creationists occupy an admirable position in a debate. A scientist who opposes them faces the same situation as a boxer battling a pair of remote-controlled boxing gloves. He can try to defend himself from punishment, but he lacks a target at which to strike back.

The above analogy holds for the central part of the Creationist doctrine, the sudden creation of the universe and its contents by supernatural means. In certain limited areas the Creationists have extended themselves, perhaps unwisely, to the defense of positions in which their credibility can be tested. In particular, they have maintained that the earth is only a few thousand years old. The earth does have an age, and its chronology can be explored by science, without any reference to the existence or nonexistence of a Creator. In an earlier chapter we described the extensive evidence, derived from the study of the behavior of radioactive elements in minerals, that supports an age of about 4.5 billion years for the earth. As a further exploration of the value of the term "science" in creation science, we shall examine the Creationist response to these findings.

The Age of Rocks Versus the Rock of Ages

"Christians desire that their children shall be taught all the sciences, but they do not want them to lose sight of the Rock of Ages while they study the age of rocks," wrote William Jennings Bryan.

Fundamentalists who viewed the age of rocks as a temptation devised by Satan must have observed the introduction of radioactive dating methods with great dismay. By the early twentieth century, their debate with evolutionists had settled into familiar lines. The authority of the Bible was pitted against the somewhat uncertain inferences drawn from the examination of accumulated sediments and fossils. The front lines had stabilized, when suddenly an entirely new force, a new source of evidence, entered the scene to upset the balance. These techniques were based on a firm theoretical and experimental discipline, the study of radioactivity and atomic processes, which provided more exact dates. Hypothetically, the new results could have supported the Creationist point of view by testifying to a young earth. In practice, they did the opposite, and put the concept of a very old earth on a firm basis indeed.

The chagrin and sense of unfairness that Creationists may have felt over these developments is expressed in an argument by Henry Morris in his book *Scientific Creationism:*

> Rocks are not dated radiometrically. Many people believe the age of rocks is determined by study of their radioactive minerals— uranium, thorium, potassium, rubidium, etc.—but this is not so. The obvious proof that this is not the way it is done is the fact that the geological column and approximate ages of all the fossil-bearing strata were all worked out long before anyone heard or thought about radioactive dating.

The interest in this statement lies in its historical and emotional content. Its logic can be demonstrated by constructing an analogy in another area: Many people believe that travelers cross the Atlantic by airplane, but this is not so. The obvious proof that this is not the way it is done is the fact that travelers crossed the Atlantic by boat long before anyone heard of airplanes.

Such words notwithstanding, great numbers of tourists now fly between Europe and America, and many rocks have their ages assigned by radioactive methods. Creationists, bound by their mythology, cannot simply bow to the new evidence but must somehow circumvent it. The simplest and most honest strategy would be a retreat to a religious position, and sometimes they do this. Again, we can cite Morris directly: "The only way we can determine the true age of the earth is for God to tell us what it is. And since he *has* told us, very plainly, in the Holy Scriptures that it is several thousand years in age, and no more, that ought to settle all basic questions of chronology."

Once such a position of faith is taken, there is no need for a believer to examine the evidence, no matter how massive it may be. If he should choose to do so, however, he could tuck it neatly within his belief system, using a principle expounded in the book *Omphalos* by Philip Henry Gosse over a century and a quarter ago. The book title is derived from the Greek word for navel, and refers to the question of whether Adam possessed one. There was no need for Adam to have a navel, as he was a product of direct creation, rather than birth. The lack of one would have set him apart from other men, however. He would be less than a man. Gosse argued that the Creator would fashion Adam as if he had a history, with hair, fingernails, and other characteristics that implied past growth.

Similarly, the earth would be created with the appearance of past existence. Rivers would flow in their beds, rocks would be weathered, sediments in place. By extending this idea, we can imagine that a Creator would also have the power to create a radiochemical record of a nonexistent past, by placing appropriate amounts of radioactive minerals, argon, and other decay products in the rocks.

Such an argument could not be tested or refuted; it would be mythology rather than science. As such it would coexist with an infinity of alternatives. For example, we could claim with equal validity that the earth and its contents (including our memories) were created ten minutes ago. This statement would not trouble the believer, who would *know* in advance which explanation was the correct one.

Creationists have attempted to pass off their doctrines as science, however, and so have set a formidable task for themselves: to confront the mountain of evidence produced by radioactive dating methods. In certain arguments, they have simply attempted to dismiss the data, on some casual pretext. A strategy of this type can have immediate impact in a debate, and serve to calm the feelings of uneasy believers. In the long run, however, it will carry little weight, as we saw in another context in the case of the Cal Tech professor described in Chapter 1.

Dr. Harold Slusher, the Creationists' authority on physics and geology, has written: "The age of the earth has had nearly as many values as the number of people who have studied the matter.... Currently, the evolutionists are maintaining with dead certainty that the average 'age' of the earth from various radiometric techniques is 4.6 billion years (within a few hundred million years). After all the claims they have made in the past which have turned out wrong, they still do this with a straight face."

Of course he is criticizing science for being science, for providing answers that are subject to revision and improvement. For those who prefer answers that are truly given with certainty and never change, religion is a much better choice.

Henry Morris takes a somewhat different tack, claiming that "no one can possibly *know* what happened before there were people to observe and record what happened. ... Scientifically speaking, no one has proof for any dates prior to the beginning of written records." Again the emphasis on *"know"* and "proof" indicates a desire for certainty. Religion, but not science, places special value on historical, written records, or at least certain selected records. Morris should have written "religiously speaking," rather than "scientifically speaking."

These dismissals are only diversions. Creationists understand that they must confront the actual evidence if they are to present themselves as scientists. They have every right to attempt this task. Scientific paradigms are not sacred; they remain open to challenge. But the evidence supporting them cannot be ignored in such a challenge. The new solution must accommodate existing data and supplement it with substantial new material, if it is to be taken seriously.

To obtain a rough estimate of the amount of data that has been obtained using radioactive dating methods, I visited the science library of my university, which has placed no special emphasis on geology. Yet there was an entire shelf of books on the subject of geochronology. One of them, a work of 250 pages published in 1969, dealt only with the potassium-argon dating method. A scientific book of this type is called a monograph. Usually, it will not contain original papers but only a review of the literature, with references to the publications which contain the data. This book contained hundreds of references, which contained many thousands of individual age determinations.

When an effort of this magnitude is directed toward any scientific technique, sources of error receive plentiful attention. Individual chapters in the monograph described different types of errors, and suggested methods by which they could be avoided or corrected. After these defects had been considered at length, it was judged that the technique could still be considered very reliable.

How could a conclusion of this sort be overturned, in the normal practice of good science? It would be necessary to attack the entire mountain of supportive evidence, one stone at a time, and disassemble it. Dr. Harold Slusher has attempted to move the first pebble. He has published a book in rebuttal, titled *Critique of Radioactive Dating* (Institute for Creation Research Technical Monograph No. 2). This "monograph," however, is only a pamphlet. It contains fifty-eight pages, only two of which are devoted to potassium-argon dating. No balanced discussion of the evidence is attempted, as in the scientific monograph in my university library, nor is new evidence presented, with supporting data. Slusher simply cites possible sources of error and presumes that they discredit the entire technique.

Perhaps he and other creation scientists have done as well as they can, considering the task they face. Imagine, for example, that we were given an equally unreasonable task: to demonstrate that Japan had triumphed over the United States in World War II. How would we go about it? First, we would have to discredit newspapers such as the *New York Times,* which contain a detailed day-by-day account of the American victory. We might first collect typographical errors in the *Times* and instances when errata were published, retracting previous mistakes. After that, we would collect a list of unsound predictions: optimistic statements by economists, prizefighters, and election campaign managers that were published in the *Times* and proved incorrect. We would put all these instances together and conclude that the *New York Times* was of no value whatever as a historical source.

We would then print an alternative newsletter that had the "authentic" information, and give its publisher a high-sounding name such as the Japanese Victory Research Institute. In it, we would publish photographs of the raid on Pearl Harbor, transcripts of Japanese wartime broadcasts that claimed imminent victory, and current news concerning the spread of Japanese cars and Japanese restaurants in the United States. Finally, we might demand that this point of view be given equal time with the conventional one in public school history classes. We could not expect to carry the day with these efforts, but it would be interesting to see the confusion we could create. Such has been the Creationist strategy in the areas they have chosen.

The guiding spirits of the movement have no illusions about the true nature of the doctrines they have represented as science. They have been remarkably candid in print. Henry Morris has written in his book *Scientific Creationism:* "Creation . . . is inaccessible to the scientific method. It is impossible to devise a scientific experiment to describe the creation process, or even to ascertain whether such a process *can* take place. The Creator does not create at the whim of a scientist."

Elsewhere Morris has written: " . . . we are completely limited to what God has seen fit to tell us, and this information is in His written Word. This is our textbook on the science of Creation!" Duane Gish has supported the same concepts in his book *Evolution: The Fossils Say No!:* "We do not know how the Creator created, what processes He used, *for He used processes which are not now operating anywhere in the natural universe.* This is why we refer to creation as special creation. We cannot discover by scientific investigation anything about the creative processes used by the Creator."

The claim that the universe, the earth, and life were made by an undetectable Creator using supernatural powers falls outside of science. It makes no predictions that can be tested. It cannot be negated by science. If it had any real possibility of negation, it would lose many of the advantages that it offers to its adherents. It is mythology serving to buttress a religion. In this sense, the use of the term "creation science" has no more meaning than would the phrase "Father Raven science." It would apply only if we wished to move the word "science" far from its accepted range. Williams Jennings Bryan wrote, just after the Scopes trial: " . . . the science of 'how to live' is the most important of all the sciences." But it is just this area of values that religion would prefer not to abandon to science.

Creationist organizer Paul Ellwanger has accepted the same central point, stating: " . . . we're not making any scientific claims for creation, but we are challenging evolution's claim to be scientific."

Our book deals with origins, and not with the details of the theory of evolution, which is concerned with life's development rather than its start. The topic has been treated well and at length by others. We must pause, though, to consider Ellwanger's comment, for it deals with the distinction between science and mythology.

The theory of evolution has all of the characteristics of a scientific statement, and is the ruling paradigm in its area. As such, it could be modified or even overthrown *if* a sufficient weight of evidence opposing it should turn up. For example, we see humans in combat with dinosaurs every day in comic strips and television films. If their coexistence in time were documented by a series of well-defined fossils, evolution would be in trouble. Alternatively, if viruses were collected in space containing messages designed for our improvement, Hoyle's theory would step to the forefront. Many routes exist for the negation of Darwin's theory. Evolution passes as science; creation science, by its own admission, does not.

The Verdict

The case of *McLean* v. *Arkansas Board of Education* was a rout. Judge Overton struck down Act 590, using such careful terms in his decision that there was little room for appeal. The decision applied only to Arkansas, but it was felt that it would influence future cases as well. The judge based his decision on various grounds, including constitutional issues and academic freedom. He developed a definition of science essentially in accord with the one we have used, and cited the Creationists' own words in deciding that creation science was not science but religion.

The trial, though one-sided, provided some interesting testimony. The Creationist side attempted to provide scientific witnesses to testify to their point of view and thus balance the flow of established scientists called by the opposition. The most reputable person that they could produce was Chandra Wickramasinghe, the collaborator with Sir Fred Hoyle in their own approach to the origin of life. He was invited, presumably, because he and Hoyle had endorsed the concept that life on earth was the product of a creator. Perhaps the Creationists were unaware that the being specified by their chosen allies was a complex silicon chip, rather than any conventional deity. Alternatively, they may have hoped to spread embarrassment and confusion among the orthodox scientists.

Wickramasinghe did affirm the view that life was the product of a creator, but spent most of his time in publicizing his specific ideas on viruses and comets. He later agreed that no rational scientist could endorse flood geology or an age for the earth of less than one million years. Judge Overton was "at a loss to understand why Dr. Wickramasinghe was called in behalf of the defendants."

This decision will not settle the deeper conflict, any more than the Scopes trial did so. A Creationist law concerning education was also passed in Louisiana, and it too was challenged by the American Civil Liberties Union. A federal judge has overturned this law as well, but his decision may be appealed. New legislation may appear in other states in the future.

The same battle is fought on a broader scale in thousands of local school board meetings, where curricula are set and text-books adopted. Local board members may come unprepared on a particular evening and be given less chance to consider the nature of science than a federal judge can demand. In these contests, the Creationists are not so much interested in advocating the practice of religion, which they can do in many other, less controversial ways, but rather are trying to subvert the practice of science in areas where the conclusions reached by scientists do not please them.

Science and religion each have their place in human affairs. Neither is served in the end by efforts to erase the distinction between them. In the origin-of-life field, the Creationists are the group that have made the most extreme attempt in this direction. Their methods include the selective citation of data, an absence of skepticism toward their doctrines, and a lack of interest in critical experiments and the concept of negation. Unfortunately, they are not alone in these practices. As we have seen, the above description also applies to adherents of many of the existing theories in the field. Mythology has penetrated to such a degree that it is difficult to judge the actual extent of our scientific knowledge.

In later chapters we will consider possible remedies for this situation in the future, but first we will pause for a final overview of this area today.

eleven

A Maiden of Doubtful Virtue

Once the spirit of skepticism has been relaxed in the major paradigm of a scientific field, it is difficult to limit the process. Variants may then appear which proclaim even more fanciful and spectacular solutions. The content of mythology increases. In the case of the origin of life, we have seen that the Creationists mark the logical end point of this process. They abandon doubt entirely in favor of the word of authority but still prefer, for tactical reasons, to call their enterprise "science."

To illustrate this idea, we have considered a number of theories in turn during the course of this book (though not always strictly in order of increasing content of mythology). In following our sequence, we have not meant to imply that each new idea replaced the previous ones within the scientific domain. Rather, they have coexisted, with each broadcasting its content independently of the others, like so many transistor radios at a public beach.

Ordinarily these "broadcasts" are segregated from one another, in separate articles, books, and conferences. On occasion, however, they are brought into close proximity at an important meeting, with the results that might be expected. During the time when I was writing this book, I had the chance to experience such an event in person.

Some 250 scientists interested in the origin of life gathered in Mainz, Germany, in July 1983 for the Seventh International Conference on that subject. This series had begun in Moscow in 1957, and more recently had settled into an every-third-year routine. This was the first that I had attended, and the number seven seemed auspicious. If, according to the Bible, only seven days had been required to create life (and mankind and the universe as well), then surely seven international conferences should be enough for scientists to unravel the mechanism. As you may have guessed, this was not the case.

The site of the meeting was more representative of the state of the field. Mainz is an old city, ravaged and devastated by many battles in its history. After World War II, it was separated administratively from its suburbs across the river, and there is no prospect of their reunion in the near future, according to a representative of the mayor's office. A similar history and state of division characterizes the origin-of-life question.

Essentially all of the opposing points of view were represented at the meeting. Representatives of the nucleic acid faction and the protein faction turned out in force, and the newest group, the clay advocates, were also quite visible. We heard of hypercycles and replicators, dust clouds in space and hot springs on earth, stromatolites, coacervates, and planetary orbiters. Sir Fred Hoyle was not present, but an astronomer described chemical evolution in space and suggested that organic material (if not bacteria) had been delivered to the early earth in comets. No participants identified themselves as Creationists, but one paper argued that both the universe and life had come into existence in a highly structured rather than simple form. Stanley Miller compared amino acid synthesis by electric discharge in reducing and nearly neutral atmospheres, and Sidney Fox spoke of the lifelike properties of microspheres. The field remained much as it had been in the past.

Occasionally there were dramatic moments as opposing viewpoints were juxtaposed with one another. Leslie Orgel presented new results from his system in which a single RNA strand was converted to a double helix, without aid from proteins. Klaus Dose, an organizer of the meeting and an outstanding exponent of the protein-first school of thought, then asked Orgel where the *first* nucleic acid strand had come from. Orgel, who speaks in an extremely candid, fluent, and concise manner, answered simply: "I have no idea how the first polynucleotide originated." Dose, paraphrasing Louis Pasteur, subsequently commented to Sidney Fox that this had been the day when a mortal blow was delivered to the nucleic acid point of view.

Another controversy concerned the origin of the preference of living systems for left-handed amino acids and right-handed sugars. Several speakers described unsuccessful experiments which attempted to assign the preference to basic physical forces operating at the atomic level. An elderly gentleman from Austria then expressed a "dissident" point of view: that the selection had been accidental. He was not content simply to advance his own solution to the problem, but commenting on the others, stated: "Unfortunately, most of the effort at its solution has gone into wrong directions." He then questioned why his own correct viewpoint had been neglected. To rectify this situation, he advocated that efforts to demonstrate the opposing theory be abandoned!

Not all of the anticipated confrontations worked out in this manner. A noted geologist, Bill Schopf of UCLA, was given forty minutes to summarize the early record of fossil life. He used a fair amount of his time to explain why Isuasphaera, from the 3.8-billion-year-old Isua rocks of Greenland, was not a fossil at all, but a mineral residue. The principal advocate of the yeastlike character of Isuasphaera, the German scientist Hans Pflug, was the next scheduled speaker, with an equal allotment of time. However, Pflug did not defend his position, but simply waved his hand in the air and said: "I will not enter into a discussion of whether these are biological organisms or not, you know my view in that respect." He spoke of other matters instead.

To even appear on the podium of the lecture hall, an ornate large room with chandeliers situated in a Renaissance palace, was an honor at this meeting. Many more participants wished to present their work than was possible in the available time. A few were selected to speak while the remainder had to be satisfied with a poster presentation. They were allotted bulletin board space in a far less elegant room downstairs, where they could hang up printed announcements of their results. The display that resulted more resembled a row of advertisements on a subway platform than science.

These displays had a certain advantage, however, in that they could remain mounted for days, while in the lecture hall the pressure for minutes was intense. Leslie Orgel, for example, who had some of the newest and most exciting results of the meeting, was given only ten minutes. In his presentation he made repeated reference to his limited time. Other speakers attempted to extend their talks by taking no notice of the chairman waving a sign announcing that their time had expired, or claiming that they had only one more slide, and then showing a series of them.

I myself could not resist the chance to put up a poster display, and mounted some signs describing "The Improbability of Prebiotic Nucleic Acid Synthesis." As I was then in the process of writing this book, I felt somewhat like a novelist who had inserted himself into his plot in the middle of the last chapter, with an opportunity to interact with and influence his characters. I was overcome by shyness, however, and spent most of my time reading other people's posters, or stood some distance apart from my own, watching those who came to read my display. As far as I could observe, only those who would agree with my presentation came over, while others whose work would be adversely affected if my ideas were correct stayed away. At the end of the meeting a younger scientist from NASA commented that I was "swimming against the tide." So much for the lethal blow to the nucleic acid point of view.

Others at the meeting took a bolder stance with regard to their own work. Clifford Matthews, a chemist from Illinois, lectured to a hushed audience with all the force and enthusiasm of a carnival barker. He had the idea that the interstellar dust clouds had been formed by the disintegration of previously existing planets. A large sign on his display asked "Where have all the planets gone?" while he, pointing first to the cosmos above and then to the ground, emphasized that the material out THERE comes from HERE. Later that evening, after an extensive tasting of German wines, a happy group of us clustered around a piano and sang Cliff's motto to the obvious tune—"Where Have All the Flowers Gone?"—while he joined in as merrily as the rest of us.

More than a quarter of a century had gone by since the first international meeting on this topic, and only a handful of the original participants were present now. One link between the first and most recent conferences was the weather. It was unusually hot in Mainz, and Stanley Miller told me that the Moscow meeting had been hotter yet. He had left the meeting hall often to escape the heat in it and obtain a cool drink outside.

I found that the temperature bothered me mostly at night, in my hotel room. There was no air conditioning, so I had to keep the windows wide open. This in turn brought in traffic noises, and the heat-and-noise combination often kept me awake. Ironically, I had little difficulty falling asleep in the lecture hall during the day, even though it was also hot and noisy. At one point I had considered sitting up in my room at night and playing the tapes I had made of the meeting during the day, in the hope of falling asleep.

The Marxist connection to the origin of life had certainly diminished from the first meeting to the present one. Only one of the papers given in Mainz made obvious reference to dialectical materialist principles, in that it related the origin of life to the evolution of higher individuals and societies. The proceedings of the first Moscow meeting had a number of references of this sort. Further, although a number of participants from the Soviet bloc were listed in the advance program of the Mainz meeting, many did not attend "for unknown circumstances." Empty poster display areas mutely advertised their unexpected absence. An elderly Soviet biologist, A. A. Krasnovsky, did preside at a number of functions, in this way maintaining the tradition started by Oparin.

This arrangement was due to unfortunate circumstances, however, rather than by plan. The Mainz meeting also served as the fourth meeting of ISSOL, the International Society for the Study of the Origin of Life. Krasnovsky was the senior ISSOL officer in attendance, as a vice-president. The president, F. Egami of Japan, had died since the last meeting. Cyril Ponnamperuma, the other vice-president and now president-elect, was ill and unable to attend. Krasnovsky thus had the honor of appearing in Oparin's place.

He was gray-haired, and wore suits with ties despite the heat. At times he appeared austere and forbidding, at other times more benign. At one reception, he accepted a gift from the mayor's office on behalf of the society, and endured the multiple mispronunciations of his name with some impatience. He then gave a speech in which he urged that science and politics should not be mixed; the origin of life was surely a topic on which scientists could find unity. I hoped to myself that we would not have to hold our breaths, or delay drinking the ample supplies of German wine present, until these wishes came true.

Krasnovsky also was nominally in charge at the ISSOL business meeting, but Americans actually ran the show. Treasurer Bill Schopf gave a careful account of the disposition of the meager finances of the organization, a few thousand dollars derived from members' dues. Most of the sum had been spent for travel fellowships to permit students to attend the meeting. Behind these small sums, however, were much larger ones which sustained the field by providing much of the support for research and for meetings of various types. The actual but barely visible sponsor was the U.S. National Aeronautics and Space Administration (NASA). The secretary of ISSOL and the editor of the society's newsletter was in fact Donald DeVincenzi, head of the Washington office that dispensed NASA funds for studies of the origin of life. Don was at the conference to present an account of NASA's own plans for space exploration relevant to the origin of life, to keep abreast of the accomplishments of those who had received NASA grant funds, and to act as a functioning officer of ISSOL, all at the same time.

The official journal of ISSOL, *Origins of Life*, is also in American hands, with Jim Ferris of Rensselaer Polytechnic Institute as editor. Ferris reported to the membership at Mainz on the state of the journal. He was asked by Krasnovsky whether the editorial board met to approve the contents of each issue, as was necessary in the USSR. Ferris replied that this practice was not needed in American journals, that the editor could act alone. His problem was a different one—to obtain enough manuscripts to fill the journal. In many important scientific journals, publication of papers is delayed because a huge backlog of manuscripts has piled up. Allen Bard, editor of the *Journal of the American Chemical Society*, once told me, for example, that about thirty submissions arrived on his desk every day. He needed two secretaries and additional telephone lines in his office at the University of Texas in order to deal with them. In the case of *Origins of Life*, however, papers were delayed in publication because not enough manuscripts had arrived to fill up a 100-page issue of the journal. In a world of science groaning under the weight of publications, this was a quiet and neglected corner.

A final item of business was the selection of the next meeting site, for 1986. NASA was to cease its role as a subdued presence and would be the host of the next meeting, at its Ames Laboratory facility in the San Francisco Bay area. It was recognized that the actual site would probably be at some nearby university, as the Ames site was fenced in and guarded for security purposes.

The final social event of the meeting was the closing dinner, at which the Oparin Medal was awarded. This award had been presented for the first time at the 1980 meeting, to Cyril Ponnamperuma. The name of the next recipient was kept secret in advance of the dinner, as if it were an Oscar award. The medal was created with the provision that it be given to the person who had made the best contribution within the past three years. This provision was deleted at the business meeting, however; it would instead be awarded for lifelong contribution to the field.

I thought it quite fitting that this be the disposition of the medal. Often, during social events at this congress, I found myself sitting next to a stranger. We would ask each other what type of science we did, and I would admit to being a biochemist, while my companion might be a geologist, an astronomer, or a microbiologist. At such times I felt as if I were visiting a shady saloon on the wrong side of the tracks, and that the drunk on the next stool and I were reassuring each other that in real life we had secure identities as a lawyer, stockbroker, or the like. We would be awed and shaken by the realization that there were some in the bar who had no other identity, and spent their entire lives in the saloon.

Some special recognition, and with none of the disrespect the above analogy might imply, certainly belonged to those willing to devote their entire career, or a large portion of it, to an area often regarded as on the fringe of respectable, hard, science. The medal was named for A. I. Oparin, perhaps the first well-known scientist to give himself fully to this field. Those who had followed his example would be the most appropriate candidates. But who would the actual winner be? This question became a favorite conversation topic at the conference.

Before the dinner, rumors circulated that it would be Sidney Fox or Stanley Miller. I had become convinced, through personal observation, that it would be Miller. I had gone to a concert with him early in the week, and we had spent a pleasant evening exchanging stories of famous scientists, but I could detect no inkling that he felt he himself would be honored in the near future. Toward the middle of the week his mood brightened visibly. At the business meeting, he came in late and sat down next to Bill Schopf. They shook hands, and then, watching his lips, I saw Stanley ask Schopf, "Do the members know?" Schopf shook his head. So for me at least, the suspense ended at that point.

The dinner itself was held in the same large and ornate room that had housed the lectures. I liked the idea that the same space which had served for confrontation could now be used for more informal encounters, over cocktails. When it came time to be seated, however, the members sorted themselves into their usual constellations. I had to make some commitment, and selected a seat at the clay table, next to Graham Cairns-Smith and his closest clay collaborator, balding, bearded Hyman Hartman of MIT. They were lost in joint plans for a forthcoming clay conference to be held in Glasgow, but I was rewarded with the company of a vivacious female scientist from NASA, who spoke of the problems of women in science.

As I had expected, Miller was selected to be honored. He was given the medal by Krasnovsky, who cited his early work on the formation of amino acids, and hoped that he would be the one to provide the answer to "the next step—the formation of the genetic code." In his acceptance speech, Stanley wisely avoided that topic and instead gave an informal, candid, and fresh account of the historical circumstances surrounding his famous experiments. He described the early negative results and his perseverance: "I wasn't interested in oil. We decided that amino acids were the most exciting thing to look for." The award and the chance to remember those times had made him for now a happy man.

Later that evening, when the courses had all been served, the speeches made, and the wine consumed, the delegates went their separate ways. They would encounter one another at other meetings in the months and years ahead, to exchange the same opinions. Those who had come secure that they had some or all of the answers to the origin of life left in the same condition. Others who had come with the doubts of the Skeptic but hoped for some convincing new answer also left as they had come. The missing fragment, the piece that would make all the others fit together, remained for the agenda of some future meeting.

Dust in the Museum

Shortly after my return from the Mainz conference, I decided to revisit one of the first origin-of-life displays that I had seen. The American Museum of Natural History in New York has carried an exhibit on this topic for the past twenty years. Several cases stood along one wall near a three-dimensional model of DNA. Within the cases were photographs of microspheres, a diagram of a Miller-Urey apparatus, an account of the prebiotic soup, and literature references for further reading. I remembered the appearance of this display, fresh, bright, and provocative, shortly after its opening in the early 1960s. It occupied the same site two decades later. The cases were filled with dust, however, and the lighting was now so dim that the words could barely be made out. The list of references still had no entry later than 1964. The neighboring DNA model, better lit, looked robust by comparison.

The sad fate of this display in a way represents the condition of the field itself. In part this comes from its close identification with the space program. In the euphoria following the Apollo moon landing project, the answers to many fundamental questions seemed close at hand. Who could tell what basic information about life would be brought back from the moon? The early returning astronauts were put into rigorous quarantine for days, to avoid infection of this planet. Even if no living organisms were to be expected on the moon's surface, moondust might contain a wealth of organic materials, perhaps even dormant spores. As for Mars, our imagination had been stimulated for years by H. G. Wells and Orson Welles, Edgar Rice Burroughs and Ray Bradbury. Perhaps we needed only to place a camera on the surface to detect everything, from exotic plants to creatures the size of polar bears.

With these exceptions in our minds, the actual reality could only be disappointing and puzzling. Enthusiasm for the space program in general, and exobiology in particular, waned, and with it, the level of funding for planetary exploration in NASA. At the height of the Apollo project, the Vice-President of the United States could advocate a manned landing on Mars by the end of this century. Later in the 1970s we were reconciled to the idea that future exploration would be conducted by unmanned spacecraft. By the early 1980s, even this less ambitious plan was threatened, and the entire planetary program seemed headed for extinction. The difficulties of that period were associated with a large reduction in federal expenditures following a change of administration. Cuts in spending for planetary exploration had started much earlier, however. Our nation today is not appreciably poorer than it was ten or twenty years ago. If it wishes to spend less on the exploration of our universe now than it did earlier, that difference must be due to a loss of aspiration, rather than impoverishment.

This loss of funding has been accompanied by a shift to pessimism concerning exobiology by some scientists, as though one development were related to the other. For example, Lynn Margulis of Boston University, then chairperson of the Planetary Biology and Chemical Evolution Committee of the National

Academy of Sciences, wrote in the magazine *The Sciences:* "There is currently no evidence that life exists anywhere else in our solar system at all." Since we will not be ready to travel to other stars for some time, "the chance for the direct detection of life beyond Earth in the near future looks very bleak indeed."

The prospect grew even bleaker with the publication of a report by the committee mentioned above which stated: "We view the search for present life in the solar system as completed: there is strong evidence that neither the planets (other than Earth) nor their satellites provide conditions consistent with the maintenance of life." If no further search is made, of course, discoveries in exobiology can hardly be expected.

Such unwarranted pessimism prepares the way for even gloomier suggestions. We have no evidence for life, of course, anywhere in the universe other than on earth. Physicist Michael Hart and others have argued on various grounds that we may be the only life, or at least the only intelligent life, anywhere. If so, then our origin may have been the result of a very unlikely event, the details of which have been lost forever, along with the early earth. By this logic, efforts to find a scientific answer to the origin of life will prove fruitless.

Gloom concerning the detection of life elsewhere can thus be extended to prospects for origin-of-life research here. The two questions have a more direct and practical connection, however. Funding for origin-of-life research in this country has largely become the province of NASA. Its influence on the entire field is profound, as more than half of the worldwide membership of ISSOL is drawn from this country.

The space agency justifies this connection with claims that "all steps in the origin and evolution of life are inextricably linked with the physical and chemical processes of cosmic evolution." This statement is certainly true in that life could not have originated on earth, or arrived here, if the earth had not been created by the processes that made the solar system. However, the specific steps in the origin of life here could well have been governed by local environmental factors on this planet and have had no relation whatever to chemical events in the interstellar dust clouds or within comets.

Many possibilities for the origin of life remain open at this point, including the cosmic link, so the interest of NASA in the question appears valid. Much odder, though, is the lack of interest by other federal agencies. According to NASA administrator Donald DeVincenzi: "The other agencies have not yet taken that direct a role in origin-of-life research. If you ask them why, they don't give a very logical answer. They just say, that's NASA's program."

Whatever the reasons involved, this concentration of power over funding seems unfortunate. Some points of view will inevitably be preferred to competing ones, and the losers have no alternative resource. Dangers of greater magnitude may also come up, and threaten funding for the field altogether.

In the summer of 1982 I met an old acquaintance, Gerry Soffen, at a conference in New Hampshire. We had both worked as research scientists in the biochemistry department of NYU more than twenty years earlier and had spent some pleasant times together. We renewed this habit, and went on a tour of the campus of Dartmouth College. During that tour, he asked me whether NASA support for the origin of life should be discontinued.

Gerry was being playful, the devil's advocate, yet I had a certain foreboding as I heard his question. Our careers had taken different paths. He was at that time Don DeVincenzi's boss, the head of all NASA biology projects. Some future person in his office might take the question quite seriously.

Financial pressures at NASA will of course affect the study of the origin of life, yet the decline of the field cannot be attributed only to the difficulties of the space program. The enthusiasm of twenty years ago drew strength from the emergence of the Oparin-Haldane paradigm. Amino acid synthesis, the preparation of adenine from hydrogen cyanide, proteinoid microspheres, all were relatively new developments. The new paradigm had rescued the field from a previous decline, after the demise of spontaneous generation. Previously, Oparin stated in 1957, many scientists felt "that it was an insoluble problem and that to work on it was unworthy of any serious investigator and was a pure waste of time." Oparin clearly felt that the new theory had reversed this situation.

At that time it seemed likely that many new spectacular syntheses, further confirming the basic concepts, might be reported at any moment. But the expected rush of new developments has not been forthcoming. Rather, doubt has been cast on two of the basic premises, the reducing atmosphere and the prebiotic soup. Scientific unity has fragmented, and very unusual ideas have moved to the center of public attention. Many workers in the field, however, have paid no attention to these changing circumstances and continue to issue optimistic bulletins. Every so often I encounter an article or press report that presumes that the major problems have been solved. Only subsidiary ones, such as the origin of the genetic code, require attention.

The effect of such publicity is simply to reinforce the credibility gap that has existed between the origin-of-life field and much of the rest of science. I have experienced this at first hand, from departmental colleagues, after telling them of my intent to get involved in the area. Their comments varied from "How can one possibly learn anything about that?" to a concerned "We don't want to lose you entirely to outer space." My personal experience is not unique. A *Nature* editorial, amidst the euphoria of the 1960s, commented: "Those who work on the origin of life must necessarily make bricks without very much straw, which goes a long way to explain why this field of study is so often regarded with deep suspicion."

This suspicion comes not only from scientists and the public in general, but from those in a related field, evolutionary biology. The author of one textbook on the subject noted that "a bias against considering origins exists in the minds of many evolutionary biologists." Presumably they wish to safeguard their field, besieged by the Creationists, from guilt by association with a less well established one. Creationists, in turn, may be trying to exploit the same connection. A 1981 article on the Creationist controversy in *The Sciences* reported: "The Creationists have been able to devise rather clever arguments against the evolutionary views of life origins. In fact they may have found and hit the Achilles' heel of modern evolutionary biology."

In this particular area, the Creationists are well qualified as critics. As a group who themselves have attempted to pass off mythology as science, they can readily identify rivals who are attempting, even if unconsciously, the same substitution. Angered that another mythology has passed muster for presentation in science classes, they cannot see why they should not have the same privilege.

Their solution, of course, takes us entirely in the wrong direction. We do not wish to convert science classes into forums where equal treatment is given to competing myths. Nor should we wish the origin-of-life field to remain in its current position within science: its reputation resembles the one, in days gone by, of a maiden of doubtful virtue, whose every appearance in public was accompanied by a background of unpleasant whispers.

How can the field be reclaimed to scientific respectability, and turned in a direction where progress can be made on important unsolved problems? Surely not by the present practices, in which prebiotic experiments are designed to gather evidence to support one point of view against the claims of its rivals. We need instead critical tests, in which one possible outcome is that an existing balloon will be punctured. The idea of negation is perhaps the scientific tool most needed in this area.

Realistically, we cannot expect those who created the existing myths to test them critically and set them aside if they do not pass muster, any more than Henry Bastian would have renounced spontaneous generation or Trofim Lysenko embraced the genetic role of DNA. The task must fall on investigators with experience in other, more exacting areas of science. Rather than ignore the origin-of-life field publicly and snicker about it privately, these scientists must be willing to apply their own stringent criteria to it, and to report back to the media when appropriate: "We don't know the answer to this."

The lack of a complete answer to important questions does not shame a field of science. This distinction is shared by many vital areas of contemporary research, such as those dealing with the cause of the aging process and the nature of consciousness. Nor does it leave us in total ignorance. We are not faced with a choice between the completed painting of a landscape and a blank wall. In the particular case of the origin of life, advances in geology, molecular biology, and astronomy have provided a frame for the picture, and negative experiments have indicated where background areas may be located. In addition, a few tantalizing brush strokes have been added here and there. Using some imagination, we can sketch in possibilities to represent the way that the completed painting might appear. Such efforts should not be made in order to create new myths. They should be labeled clearly as speculations: suggestions that are consistent with existing evidence but go well beyond it and offer new explanations not supported by scientific data.

Speculations can be useful, even vital, in science because they suggest new experiments and new directions for research. Those who make them have a responsibility to be clear about their nature, however, and to make suggestions for negating them as well as affirming them. With this caution in mind, I will use the remaining chapters to describe some additional possibilities for the development of life on earth, and suggest possible experiments and explorations that may bring us closer to the ultimate answer concerning its origin.

12

The Case for the Chicken

Earlier we reviewed the intense controversy that concerned whether nucleic acids or proteins had priority in the origin of life. We compared it to the debate over which came first, the chicken or the egg. After some consideration, we eliminated the egg. I mean by this the hereditary substance of today, nucleic acids. Nobel laureate Joshua Lederberg had noted as early as 1960: "There is some controversy over whether nucleic acids were the first genes, partly because they are so complex, partly because their perfection hints at an interval of chemical evolution rather than one master stroke." These words are equally compelling now. Even the building blocks of nucleic acids, the nucleotides, are intricate molecules containing over thirty atoms each and requiring the precise connection of three subunits, with the release of two molecules of water. It is not surprising that prebiotic syntheses of nucleotides have run into intractable problems. These substances probably were developed well after life began.

If we discard the egg, one answer remains to the riddle. It is time to present the case for the chicken. I shall argue the following points: (1) A hereditary system based on protein preceded those based on nucleic acids. (2) RNA was first developed as a building material, a structural support in protein synthesis. It gradually took over its hereditary role. (3) At a later stage DNA in turn evolved, and became the genetic substance. This development was related to the emergence of eukaryotic cells somewhat more than a billion years ago, and helped trigger the explosive increase in the rate of evolution since that time. Thus, the well-known Central Dogma of molecular biology, "DNA makes RNA makes protein," was exactly reversed in the development of life: In the beginning there was protein. Protein begat RNA, and then both begat DNA.

We must dismiss the Skeptic for a while, before we continue. He has accompanied us as we successively analyzed spontaneous generation, the Oparin-Haldane hypothesis, the naked gene, Hoyle's ideas, and Creationism. It is time now that we went beyond experiment and generated our own speculations. He is not adept at this process, and might even hinder it, so we will spare him.

Any discussion of life without a nucleic acid hereditary system must be speculation at this time. The only life we know is the kind that lives on earth today, and all of it uses nucleic acids in this way. We can only guess how life might work without DNA and RNA. While in principle only our imaginations need limit us, we need to use restraint in practice lest we generate science fiction rather than plausible science. For this reason, we will limit the number of new assumptions that we make, and work within the existing framework of science as much as possible.

Life Without Nucleic Acids

To start, we will assume that proteins transmitted their own heredity before they devised nucleic acids as an improved mechanism for that purpose. A number of scientists have explored this idea and suggested schemes for protein replication similar to that employed by nucleic acids. An amino acid in solution would in some manner directly pair with a partner on a protein chain, so that the sequence of amino acids on the chain under construction was controlled by the existing one. The replication of DNA works in this way, of course, according to the Watson-Crick base-pairing rules. Some suggestions have been made concerning possible direct recognition schemes for amino acids, but no convincing demonstration has appeared, although one should be possible if such a scheme existed. But perhaps the answer lies in another direction.

If proteins could replicate directly, by some simple pairing scheme, there would have been no need for them to turn this function over to nucleic acids. Proteins store the same information more economically, using less material. For example, an average amino acid in a protein chain contains about 16 atoms. The same information, stored in three units of an RNA chain, requires about 100 atoms. In DNA, the identical information is kept in a complex of two chains and needs 200 atoms. This extra expenditure of material in storing the same information would be justified only if there was a corresponding increase in efficiency in moving to the more complex systems. We must presume that the earlier protein-based hereditary system was more cumbersome, and less elegant, than the present one.

How can we model this earlier, clumsier system? For inspiration, we can look at the mechanisms that exist today. When a cell makes proteins, the information in DNA is transmitted first to RNA. This transfer is performed efficiently, using Watson-Crick base pairs. The message, still written in the language of nucleic acids, must then be translated into that of protein. Much research in molecular biology over the past decades has been devoted to the study of the translation machinery.

A search has been conducted for some direct molecular fit between an amino acid and a group of nucleotides to provide a logical connection between the two languages. If some natural pairing scheme existed, it would explain the basis of the genetic code today, and suggest events that took place when the code was first developed. No such direct fit or pairing scheme has been found, however, although a number of interesting suggestions have been put forward. The RNA-to-protein connection is made instead in a cumbersome manner.

A group of molecules exist whose technical name is "aminoacyl tRNA synthetases," but we will call them the interpreters. They are special enzymes, best visualized as two-handed molecules. Each of them is capable of recognizing and selecting a single amino acid out of the set of twenty, using one "hand." With the other "hand" it seizes the appropriate small RNA molecule (a transfer RNA) from the mixture present in the cell. The enzyme then joins the two together. The set of interpreter molecules has the responsibility for ensuring that the orders originally stored in DNA are correctly carried out in the construction of a protein.

A simple analogy may help to make this process clear. Imagine a group of human interpreters given the task of translating Chinese into English. Each individual, however, knows only one Chinese character and its English-language equivalent. The message to be translated is posted on a wall, one character at a time. As each character appears, the appropriate interpreter steps forward and places the English word alongside it. Eventually, the entire message will be translated, provided an interpreter is present for each character shown. The biological system functions in the same manner. Fortunately, the number of characters to be translated is quite limited.

The same system could work, in a simpler way, to copy protein into protein. Again, we would need a set of two-handed interpreter molecules. Their job would be easier, however, as the enzyme would only need to recognize an amino acid bound within a protein chain and the same amino acid in a free state.

The molecule to be copied would be attached to some suitable support (perhaps another protein or a polysaccharide) to distinguish it from the other proteins in the cell and mark it for duplication. While on this framework, it would be held and turned in some manner, so that one amino acid after another, in its chain, was exposed successively. Each exposed amino acid would be recognized by a specialized interpreter molecule. This enzyme would then select the same amino acid from solution and insert it into the corresponding place of the new protein chain that was under construction. With the completion of this task, a gene would have been duplicated, and in addition, another useful molecule would have been made, as every protein in the cell would both perform a function and carry its own heredity.

If this mechanism preceded the development of nucleic acids and was later replaced, then presumably it worked less well as a genetic system. It may have been inaccurate, or slow, or defective in other ways. This very inefficiency, however, suggests a solution to one puzzle concerning the pace of evolution.

To appreciate the problem, we should first review the current paradigm, shared by many scientists, which describes the development of life. It assumes that protein, RNA, and DNA all date back to the earliest days of life on this planet, 3.5 billion years ago. The fossil stromatolytes and other remains from that ancient time have shapes similar to those of contemporary organisms. By analogy, it is presumed that the internal processes of these early cells were similar to those present in prokaryotes today. If this was so, then life essentially marked time for over 2 billion years, with little evolutionary progress, except perhaps for the development of oxygen-releasing photosynthesis.

Then at some time, a billion or a billion and a half years ago, there was a rash of new developments. Eukaryotic cells arose from simpler ones, sexual mechanisms developed, and multicelled creatures came into existence. All of the larger life forms familiar to us arose within the last half-billion years. No consensus exists concerning the reasons for this belated series of events.

Much can be explained, however, if we assume that the genetic function of nucleic acids developed later in evolution, and was preceded by a cruder system based on protein. The idea of a late origin for DNA has been put forward by a number of scientists, including biologist John Keosian and physicist Freeman Dyson.

If nucleic acids came later, then the earliest fossils would represent organisms that functioned with a protein-based genetic system. The period from 3.5 to about 1.5 billion years ago, when little seemed to happen in terms of the external forms of organisms, was a period of gradual evolution under the protein genetic system, culminating with the turnover of this function first to RNA, and then to DNA. With DNA finally in place as the ultimate hereditary material, rapid evolution could then take place. The rest of the story is a familiar one, and leads in the end to our own appearance.

What were the major developments in biochemistry during the reign of protein? It is harder to tell of these events than of the rise and fall of ministers in kingdoms that flourished before the development of writing. Logic is our principal guide.

During the long, slow period of protein evolution, the number of amino acids employed may have increased from a handful to the twenty we know today. The current set varies from ten to twenty-eight atoms in size. The two smallest are prominent in Miller-Urey experiments. They were likely to be present in the first set. The largest ones are inaccessible even to the most contrived prebiotic simulations. They most probably were introduced after some development of metabolism had occurred. Some writers have suggested that a set of six amino acids would suffice to approximate the various shapes we see in proteins today. Others would reduce the initial number to four. Whatever the starting point, each new introduction of an amino acid may have marked a milestone in the evolutionary struggle of early life.

The size and sophistication of enzymes undoubtedly increased over these 2 billion years of evolution. At present, they range from perhaps 100 amino acids in a chain to beyond 1,000. These mammoth sizes permit exquisite perfection of combined catalytic and regulatory abilities. But what was the starting point for this development? There is no easy answer.

Even isolated amino acids, unconnected to others, can show modest activity as catalysts. This property does not belong to amino acids alone, but is shared by other chemicals, both organic and inorganic ones. One enjoyable activity for chemists is to design other molecules that display enzymelike properties. Some impressive results have been obtained using a doughnut-shaped carbohydrate, for example.

If we return to amino acids, however, it may be fair to talk of enzyme activity when the catalytic power of a number of units connected together greatly exceeds that of a mixture of the same units not attached to one another. Such activity begins when an amino acid chain attains the size needed to fold into a special well-defined three-dimensional shape. This may require several dozen amino acids. The era of protein-based life on earth was, then, a time in which enzymes increased from the minimal size to dimensions resembling those we observe today.

The Entry of RNA and DNA

Evolution does not anticipate needs. It is unlikely that nucleic acids were developed with the hope that they would take over the genetic function at some suitable future date. They were probably utilized for some other purpose, and gradually moved to their present position in life. In the earliest days on earth, phosphate was locked in an insoluble form within volcanic rocks and only gradually became available as the rocks eroded. When it was rare, it was probably reserved for its present role in energy storage, though simpler molecules than ATP were undoubtedly involved.

Gradually, as phosphate supplies increased, new roles were found for it. If we examine a bacterial cell today, we may note the presence of substances called teichoic acids, which have a family resemblance to nucleic acids. They contain an alternating sugar-phosphate backbone but lack bases, substituting in their place an amino acid or an additional sugar. These substances are present in the cell walls and membranes of certain bacteria. Their presence testifies that they have useful properties as construction materials, and perhaps serve other purposes as well. Many variations on the teichoic acid theme may have developed in the course of evolution. During this process, the first active nucleotide subunits were made.

The union of nucleotide subunits to form the first RNA has been a headache for prebiotic chemists. This step need not be difficult, however, when an appropriate enzyme is present. We have seen that $Q\beta$ replicase can assemble an RNA molecule on its own, given the proper subunits. The first nucleic acid may have been put together by a less specialized enzyme, with a general ability to connect phosphates and sugars.

Once formed, these brand-new substances no doubt quickly demonstrated their ability as structural materials. In fact they are still primarily used for that purpose today in the ribosome; more nucleic acid is employed for ribosome construction than for all other purposes. This usefulness derives from the same property that makes nucleic acids valuable in heredity: the formation of Watson-Crick base pairs. When a nucleic acid chain has no appropriate partner with which to form a double helix, it folds back upon itself and assumes a shape that will permit many internal base pairs to be formed. The exact form that the molecule takes depends on the detailed order of bases within it. This property may have made RNA an ideal support for the process of protein synthesis, directed by proteins. It was adapted for this purpose after its first discovery, presumably displacing some related but less suitable substance.

Once in place it could be improved, and evolve. It would be advantageous for the cell to develop better base sequences which provided more useful shapes. At first, enzymes might prepare the sequences on their own, in a cumbersome process. After a time, however, the discovery would be made that RNA molecules could be copied directly, as $Q\beta$ replicase copies $Q\beta$ RNA. Errors might occur during the copying process, and if favorable, they would be perpetuated by natural selection. This development undoubtedly led to a rash of improvements in the apparatus for protein synthesis, and a much more complex ribosome resulted, made largely of RNA.

One likely improvement was the preparation of short, specialized units of RNA, each associated with a particular amino acid. These small RNAs aided the insertion of the amino acids into the protein under construction. At this point, the vital interpreter enzyme recognized the free amino acid, the amino acid within the protein to be copied, and the small helper RNA (the predecessor of transfer RNA). As an additional aid, a longer RNA was developed that aligned, by base pairing, the various helper RNA molecules in an order appropriate for the protein to be copied. One long RNA of this type (the predecessor of today's messenger RNA) was prepared for every useful protein in the cell. With this innovation, however, the information present in each protein was also stored in RNA. A duplicate genetic system, one capable of separate evolution, had evolved.

This system proved to have many advantages. It was no longer necessary, for example, that a cell keep at least one copy of every enzyme present at all times, lest the information be lost. Enzyme levels could be raised quickly, or reduced to zero, as environmental circumstances required. Eventually, as the RNA heredity system proved effective, the now superfluous one based on protein could be discarded. Proteins were now free to perform the functions that they did best.

At this stage of evolution, the genetic information of cells was stored in a set of separate RNA molecules, each corresponding to a protein. These molecules also served the purpose that messenger RNA molecules do today: they participated directly in the construction of proteins in the ribosomes. Now, of course, these two functions are separated, with DNA acting as the remote repository of genetic instructions and messenger RNA acting only as a transient intermediate. At some point DNA was created, by the introduction of minor modifications into RNA, and hereditary information was transferred to it. This transfer of information from RNA to DNA was the reverse of the direction usually used in biology, but it still takes place in the life cycle of certain viruses, and occasionally in higher organisms, today.

It seems likely that this final genetic innovation took place at the point in evolution when eukaryotes and prokaryotes diverged from one another, perhaps 1.2 or 1.4 billion years ago. If this was the case, then another evolutionary puzzle would be solved. We have seen that eukaryotes have their coding DNA broken into segments, in most genes. A run of bases that carries the information for part of a protein gives way to a "commercial break," or intron. Then the coding portion resumes. Many such interruptions may occur before the genetic message is completed. Prokaryotes generally lack such extraneous insertions in their genes. The evolutionists who believe that DNA has existed since the earliest days of life are at a loss to explain why introns were inserted into the continuous messages of prokaryotes when they evolved into eukaryotes. This dilemma vanishes if we presume that the selection of these two different forms of organization for DNA was made soon after this molecule first appeared, and that the choice was one of the critical steps that sent eukaryotes and prokaryotes on their separate ways. Thus, after eons of innovation and changeover, our biochemical system came to its final form.

The Central Dogma

We have speculated enough, for now. It is time to let the Skeptic return. He inquires at once about the contradiction between our scheme and the Central Dogma. Should this revered theory be so lightly discarded?

Even the sound of these ominous words, let alone the capital letters, should suffice to intimidate any casual speculator who would tamper with it. My dictionary defines "dogma" as "a doctrine, belief, body of theological doctrines strictly adhered to." This term, in science, would surely suggest a theory with the most formidable support. Many texts certainly do support that impression. For example, the introduction by Richard E. Leakey to an illustrated version of Darwin's *Origin of Species* states: "Genetic information itself, flows in only one direction: from DNA outwards. The statement is called the Central Dogma of molecular genetics. It has been elaborated from a vast array of experimental data and seems unlikely ever to be seriously challenged." The dogma itself was first pronounced by Francis Crick in 1958. In its exact words, it stated: "The transfer of information from nucleic acid to nucleic acid, or from nucleic acid to protein may be possible, but transfer from protein to protein, or from protein to nucleic acid is impossible."

Thus the discovery of information flow from RNA to DNA noted above was by no means forbidden; however, the other transfers we have discussed would appear to be ruled out. But how had Crick come to this conclusion? In a later article he pointed out that it was simply a negative hypothesis. No trace of machinery for transfer of information from protein to nucleic acids had been found in contemporary organisms. That did not mean that such machinery could not have existed in the past. Crick stated explicitly in his later paper that the dogma "was intended to apply only to present-day organisms, and not to events in the remote past, such as the origin of life or the origin of the code."

To get a better picture of the circumstances that led him to form his theory, and to give it its imposing name, I went to talk with Francis Crick. He was attending a conference on the nervous system at Cold Spring Harbor, New York, not far from my home, and was happy to reminisce about the Dogma, late on a May evening. He is a striking man, tall, gray-haired, hospitable, relaxed, and above all, very merry.

He recalled that at one point after he had put forth his theory, a friend had told him that a dogma is something that cannot possibly be doubted. "I didn't know it meant that," said Crick. "I thought it meant a hypothesis, some arbitrary thing which was laid down for no particularly good reason. Otherwise it would have been called the Central Hypothesis, and then nobody would have made all this fuss."

Thus the Dogma is simply a convenient organizing idea with a misleading name. In terms of the origin of life, Crick was willing to admit that the idea of nucleic acids first raised some difficulties, and perhaps the thought of proteins first should be taken up again. He might make some new effort, perhaps with Leslie Orgel, to think about it again.

After the comment, I could not resist testing Leslie Orgel's reaction to the speculative protein-based scheme sketched above. He listened patiently as we had breakfast together a few weeks later, then said abruptly: "Enzymes can do anything."

That phrase was one I had first heard as an undergraduate biochemistry student in the 1950s. We had learned during the previous year in organic chemistry class that certain chemical reactions would work, and others would fail, according to an empirical scheme discovered by chemists at the cost of a lot of hard work. We then went to biochemistry and saw that the most unlikely reactions took place in living systems. It was only necessary to write the name of an enzyme over the chemical equation to validate the process. The then-mysterious powers of enzymes would take care of the details.

Orgel had meant to indicate that enzymes could undoubtedly carry out, in principle, the type of self-replication scheme I had described. But that did not prove that such a scheme had ever existed, and been a factor in the development of life. He commented: "Making models is just too easy. I don't feel terribly sympathetic to speculations not carrying good experiments."

That is the point, of course. Suggestions are not enough. They must have a predictive value, and lead to critical tests. But how would one set about to prove the existence of such a system, or to disprove it?

One approach would be to construct such a system in the laboratory. But current technology is not up to the challenge. We do not as yet know enough about enzymes to design even one that would perform a completely new function in an efficient manner. It would be harder yet to construct a system of interacting enzymes and get it to function. Someday this feat may be possible. We may be able to assemble from scratch a group of cooperating enzymes that would in effect constitute a simple model of life.

But such a feat would not prove our point. It would demonstrate that we had gained skill at manipulating molecules, but not that our system, however effectively it might operate, necessarily had a role in the development of life on earth. Our best chance at establishing this historical point comes from the examination of living systems as they operate today. We are fortunate to be alive at a time when exhilarating progress is being made in this endeavor. The intense study of the molecular basis of life is likely to continue, and expand greatly in the future. From this work, we will learn much about the history of life on our planet.

In the 1970s, for example, vastly improved techniques were developed to determine the sequence of bases in a segment of DNA. In 1970 workers struggled to determine the order of 20 bases in a row. A dozen years later the arrangement of 48,502 bases in the DNA of a bacteriophage called lambda was known. When another dozen years or so have passed, we can anticipate that most or all of the sequence of 4 million bases in one strand of the chromosome of the bacteria E. coli will have been deciphered. Many sequences of importance in higher organisms, including man, will be known as well.

Even at the present time, a detailed comparison of sequences of amino acids in proteins, and bases in RNA, from different organisms has been used to determine degrees of relatedness and to construct evolutionary family trees. Much more information of this type is available in DNA sequences. Ultimately we shall gain a clear picture of the order of divergence of fundamental groups such as eukaryotes, archaebacteria, and ordinary bacteria from one another.

The full meaning of the information present in DNA sequences will not be apparent immediately, but eventually it will yield to patient investigation. We will learn the full genetic capability of a bacterium, the composition of every protein, RNA, and other molecule that it can make. Physical studies with improved instruments will tell us of the way these molecules fit together in three dimensions, and we will have the complete blueprint of a bacterium.

Our knowledge of the way that an organism functions may allow us to infer details of its past evolutionary development. I have speculated that the present bacterial ribosome was derived from an earlier version that functioned with a protein-based hereditary system. Perhaps remnants of the earlier structure can be detected in the present one, just as older elements of a cathedral can be deduced by study of its present form. Similarly, the "interpreter" enzymes that we discussed may show vestiges of an ability to recognize amino acid units within proteins. The search for such remnants represents a type of molecular archeology. Relics of the past may also be discovered within the DNA sequences themselves. The DNA of eukaryotes apparently carries a number of "dead genes." Such sequences resemble working genes but have suffered some adverse change by mutation. They no longer serve to produce a protein, but are carried along in the heredity of the organism just the same, and serve as a record of some ancient accident.

Much more dramatic and to the point would be the discovery of living relics, survivors of the original protein-based system that are alive and functioning on our planet today. Microbiologists often deny the possibility of such a discovery, maintaining that such creatures would already be known, if they existed at all. But this need not be the case

The microbial kingdom of earth has been only partly explored for its content of novel organisms. Microbiologists often use a set of reliable culture media repeatedly, since these allow certain strains to multiply readily to the point where it is convenient to study them. Truly exotic organisms, however, may not grow well in the familiar media, and may escape detection, even though they are present in common environments. Others may not be distributed commonly but may lurk in unfamiliar niches on our planet, where their presence has not been suspected.

The cold, dry, windswept valleys of Antarctica, for example, were once considered to be barren of life. The most obvious places in these valleys, such as the exposed surfaces of the terrain, were in fact lifeless. But an entire miniature ecosystem of fairly conventional algae and bacteria was later found to be present, nestled snugly within certain porous rocks.

Unusual habitats might also hold unconventional organisms. A five-year project, announced by the government of Japan in 1984, has the intention of screening bizarre environments for "superbugs" with new and possibly useful properties. *Nature* magazine commented: "There is plenty of reason to believe that there are vast numbers of microorganisms living in extreme environments that are still waiting to be discovered."

Methanogens, as we have seen, are oxygen-sensitive microbes that obtain energy by combining organic material with hydrogen, rather than oxygen. They inhabit locations such as the mud at the bottom of San Francisco Bay. Twenty years ago, before their discovery, a speculative book on science guessed at the existence of such creatures, but placed them on a remote planet, far out in the galaxy.

Other surprises have come quite recently. Biologists have long believed that living things could not survive if held at temperatures above $100\,^{\circ}C$ ($212\,^{\circ}F$). Yet, as we noted earlier, microorganisms have been reported within hot vents at the sea bottom that thrive at temperatures in excess of $250\,^{\circ}C$ ($482\,^{\circ}F$). The reports have been questioned, and controversy exists as to the reality of these creatures. If they exist, then they must use novel mechanisms to maintain the stability of their important biochemicals. Whatever the outcome, we must not become too secure in our feeling that we know everything that may be alive on earth.

Somewhere on this planet, perhaps in localities depleted in phosphate, survivors from the era of protein life may yet persist, requiring only exploration and suitable culture media for their detection. Joshua Lederberg has suggested that such organisms be raised in the presence of radioactive phosphate. Conventional creatures would build this phosphate into their nucleic acids and perish when it disintegrated, while protein-based organisms would be spared.

It is possible, in fact, that the first relic of protein life has already been uncovered—the protein analog of a virus. Scientists have worked for decades to purify a particle called scrapie, which is responsible for an infectious disease that causes brain lesions and death in sheep and goats. The course of the disease requires years, and the isolation and identification of the particle has proceeded slowly for this reason.

The scrapie agent, viral-sized, appears to consist of a protein alone, with no nucleic acid. Yet it exists in different strains, and apparently has a gene, capable of mutations. How, then, does it transmit its heredity?

Perhaps a cleverly concealed nucleic acid lies within the scrapie particle. If not, some other mundane possibility may yet explain its action. Another alternative, however, is that it has a protein genome—one that codes for protein, or feeds information into the nucleic acid system of the host cell, by a mechanism that violates the Central Dogma. Such a discovery would be revolutionary. It would not only establish the possibility of a protein-based hereditary system, but also demonstrate that normal cells retain the capacity to interact with such a system.

A discovery of this magnitude would require a great deal of verification, so we need not leap to firm conclusions at this time. The discovery of existing protein-based life, however, would strongly support the idea that during the course of evolution a system of this type preceded the one based on nucleic acids.

Let us suppose for a moment that this idea is true. Would we have solved fully the question of the origin of life? Unfortunately, no. The resolution of the chicken-or-the-egg paradox would provide a picture of the development of life, back to its early days on this planet, but it would not take us to the very beginning. We would simply have returned the origin-of-life question back to a form it had in earlier days, when Darwin and then

Troland suggested that life began with the appearance of the first functional enzyme or protein. We would not know what preceded the first protein replicator.

Certain headaches are removed, of course, by the speculative scheme in this chapter. Nucleotides need not be made prebiotically. The development of the genetic code and the relation of nucleic acid to protein are postponed to a later time in evolution. Such complexities need not be dealt with in a prebiotic soup. Since amino acids are made in Miller-Urey experiments, we are in much less difficulty when we consider the availability of building blocks for the replicator. One burning question remains: How did the appropriate subunits come together to form the first self-reproducing system?

Once again, the often-abolished concept of spontaneous generation reenters to solve the problem. But once again, it will fail.

Let us imagine the simplest form of the system described in this chapter. We need a collection of small enzymes. Perhaps we can get by with only four amino acids. One enzyme will be needed to control the entry of each amino acid into a protein that is under construction. Others might serve to provide a framework for protein synthesis, to help in making amino acids, and to procure a supply of energy. A community of at least ten different enzymes would probably be required.

How complex need each enzyme be? It is hard to imagine how the necessary specificity and reaction speed could be obtained with less than perhaps 25 amino acids in each. Thus we have specified 250 amino acids in building our replicating system.

If we had to await the construction of this community by random selection from a pool containing only the four amino acids, the odds of getting it right in a single try would be 1 in 10^{150}. Of course, a number of different solutions might work to afford a viable replicator. But on the other hand, no pool that was likely to exist on the early earth would contain only the subunits we desired. Mirror-image forms of the amino acids would be there, nonbiological amino acids, and many other substances which are not amino acids at all, but are nevertheless capable of intruding into a protein chain and fouling it up. By any calculation, the odds against the spontaneous generation of a protein replicator, while much better than those against a nucleic acid replicator, still greatly exceed the number of tries available on the early earth.

We discussed another alternative earlier. Sidney Fox and others hold that amino acids do not combine at random, but according to rules inherent in their chemistry. Most chemists would agree with this, but not with the further assumption that these rules would favor the rapid formation of a self-replicating, evolving system. A marvelous circumstance of this sort must be demonstrated by exact experiments, not merely proclaimed.

Our efforts to reason backward from a knowledge of modern biochemistry to the origin of life have generated some provocative speculations, but did not take us back to the very beginning. This was no surprise, as we would not expect them to inform us about the nature of the simplest organized systems. We can understand this best through an illustration.

Suppose we wished to learn something of the adventures and hardships of the first human inhabitants of North America; not the European colonists, but rather the first migratory tribesmen who arrived in prehistoric times. We would gain little by an intensive study of the U.S. Constitution, or even the charters of the colonies that preceded the United States. Some better approximation might be made by inducing volunteers to settle a comparable uninhabited area without the benefit of modern tools. We would at least learn something of the natural obstacles to survival, and the problems to be faced in organizing a small community.

Most likely, no historical trace remains of the first chemical steps involved in the origin of life on earth. We cannot study these specific events. By laboratory simulation, however, we can explore the general principles involved in chemical self-organization. When these are understood, we can understand which variants of the process are more likely to lead in the direction of our particular biochemistry. Sidney Fox has termed this approach the "constructionist" one. The alternative, the study of existing organisms, is called "reductionist." He has compared an effort to understand origins by the reductionist approach to an attempt to learn how to bake a cake by unbaking a finished one.

Many workers, of course, have shared this philosophy. The scientific literature of the origin of life has not suffered from a dearth of prebiotic experiments. However, the vast majority of them have been conducted to achieve a desired synthetic result, rather than to search for the missing principle that governs gradual chemical evolution.

Creationists and kindred spirits have been quite astute in recognizing the defects in this approach, and have selected a religious solution to the problem. But the scientific options are far from exhausted. With a new skeptical spirit of inquiry, one free from preconceptions about the earliest events in the organization of life, we may yet penetrate the mystery. We will consider these possibilities in the next chapter.

thirteen

The Way to the Answer

At one point in the history of this planet, the simplest systems capable of producing offspring and evolving further first emerged. Fossils of one-celled organisms dating back 3.5 billion years testify to that event, which may have taken place earlier yet. The first self-reproducing systems need not even have been arranged into cells, but may have kept themselves together in some simpler way. Such beings, though far less complex than a modern bacterium, would yet be much more organized than the simple chemical mixtures from which they presumably arose. We do not understand how this gap in organization was closed, and this remains the most crucial unsolved problem concerning the origin of life.

One tradition present in mythology maintains that the gap was never closed. The organization present in life today was introduced from above, rather than below, through the act of an even more organized supernatural being. As we saw earlier, another view, that life organized itself from a primeval state of chaos, is also represented in mythology (it is reflected as well in dialectical materialism, which maintains that the continuation of this process has now produced advanced societies of a particular kind). The latter point of view of the origin of life, but not the former one, is accessible to the scientific approach and can be tested by experiment.

Many experiments have in fact been conducted, and we described them in earlier chapters. We also discussed the flaws that made them unsatisfactory as an answer to the problem of self-organization. In these cases, the experimenter had a preconceived goal. He or she wished to demonstrate the efficient synthesis of amino acids, a polynucleotide, or other biochemical entity important in life today under conditions considered plausible for the early earth. The ingredients and conditions were selected in order to maximize the likelihood of the desired outcome. Results which were unrelated to our current biochemistry were considered uninteresting—for example, the formation of oil in the first Miller-Urey experiment. They were not followed up. Rather, a different set of conditions was tried, to get the intended result.

When success in a particular step was achieved, this part of the problem was considered solved; attention could then be turned to other matters. For example, a complete prebiotic synthesis of a replicating nucleic acid is needed for the naked gene theory. In this quest, the synthesis of adenine and the conversion of a single-stranded nucleic acid to double-stranded form, without enzymes, are completed achievements. Further improvements are always welcome, of course, but not essential. An efficient nucleoside synthesis, and the further replication of a double-stranded nucleic acid, on the other hand, are steps that are yet to be worked out.

Many adherents of prebiotic theories regard such uncompleted steps simply as tiresome chores to be attended to, akin to messy rooms in a house that is being renovated. Allen Schwartz, of the University of Nijmegen in the Netherlands, said to me, for example, that he has "almost a kind of faith" that the missing steps will be demonstrated. Others show even less patience; they are willing to *assume* that the necessary work will eventually be done. This was expressed in a sentence in the report of a NASA advisory panel: "Many people feel that the efficient production of nucleosides is something that will be demonstrated sooner or later, and that it is not really on the forefront of research anymore."

This approach, unfortunately, does not represent science, but rather a search for evidence in support of an established mythology. Any scientific approach to the naked gene or other detailed theory of the origin of life must include a determined effort at its negation. A record of failure in an indispensable step such as nucleoside synthesis would provide a clear indication that the theory was incorrect. A negative inference of this type could never be fully conclusive, of course. A redeeming counterexample might always turn up to save the day. For example, some recipe may yet be found which converts simple chemicals directly into a bacterium, refuting Louis Pasteur and redeeming spontaneous generation. Lacking such an experiment, we can at least provisionally put spontaneous generation (and perhaps, after more work, the naked gene theory as well) into the trash barrel.

In theories that consist of demonstrations without any effort at negation, the experimenter controls the events, as Dr. Midas controlled Charlie the Chimp at the typewriter. There was no way that Charlie could type some other message. In most prebiotic simulations, there is no way for some other conclusion about the origin of life to emerge.

An almost uncharted area remains open to skilled chemical investigation: undirected prebiotic experiments. Some aspects of a study of this type have been anticipated. Isolated individuals have called for experiments that accurately simulate the complexity of a primitive earth environment. In 1963, at the Second International Conference in Florida, physicist H. H. Pattee made this point: "For all the inevitable inaccuracies in detail, a sterile simulated seashore, with waves, tides, sand, rain, and intermittent sunlight, is a more accurate primitive earth environment than the well-defined but oversimplified reactions studied so far." And chemist David Usher, of Cornell, has planned a "day-and-night machine," but has not as yet constructed it.

The virtue of such elaborate devices is not only that they would simulate an authentic environment, but that they would be less susceptible to the bias of the experimenter. Ideally, a study would start with the introduction of a realistic and simple mixture of chemicals into the machine. The apparatus would be turned on and let to run indefinitely without further interference by the scientists, except perhaps for the removal of a small sample from time to time for analysis.

What result would constitute a failure? The failure of any particular chemical to appear, however important it may be in life today, would not be significant. The experiment would end when the further input of energy and the passage of time produced no further significant change in the chemical mixture within the machine. This might happen at the very start, as in the case of sunlight irradiating a junkyard. Alternatively, the entire apparatus might become coated with a horrible, intractible tar, which would remain as tar thereafter. In either case, chemical evolution would have come to an end.

Perhaps we should not go to a large night-and-day machine in the earliest studies. The prospect of cleaning such an enclosure of tar is most unpleasant. The first efforts could be carried out on a small scale. The most important point would be that the scientist not interfere until the end point is reached; the size of the endeavor is a less critical feature.

Early experiments would most likely produce numerous ignoble failures, and test the patience of the investigators. But perhaps one day, a mixture would not grind to a halt or turn to tar. Cycles of chemical reactions would be set up which would persist and slowly gain in complexity. Even if they damped out after some time, we would have learned from the experience. A revised attempt could then be made.

One day, with the right mixture and conditions, the process might not end. The chemical system would slowly organize itself and continue to evolve. Initially, it might not contain the chemicals important to our biochemistry. These substances might appear later, or not at all. Either way, the result would be a vital one. By intensive study of such an evolving system, we would learn how matter can organize itself, even if the direction taken is different from the one that occurred on this planet. Once the principle was understood, the particular variation that leads to our own biochemistry could be sought, with a greater chance of success.

Chemical Evolution in the Solar System

Due to human limitations, experiments of the above type must be limited in scope and time, and may be subject to unconscious experimental bias. Places exist, however, where studies in chemical evolution have been carried out on a grand scale, without bias, for billions of years. The results await us; we need only to collect and analyze them. The answers may be stunning. Unfortunately, the collection process will be expensive, for the places are the other worlds of our solar system.

These worlds offer a dazzling selection of varied chemical circumstances. Temperatures can be hotter than our hottest sea vent or colder than a blizzard in Antarctica. We can explore solid or liquid phases, and thick, thin, or nonexistent atmospheres. Do we wish an oxidizing, neutral, or reducing environment? We need only make our selection. Further, each world has had the same length of time as the earth to work out its own destiny.

Mankind has only had the resources thus far to visit one other world, the moon, in person. It was the most convenient and least expensive one, but one of the least interesting in terms of chemical evolution. This visit represents the equivalent of a vacation on Staten Island for a resident of Manhattan. The surface of the moon lacks liquid or an atmosphere, important aids to the evolutionary process.

Fortunately, our choices do not end here. Just as our vacationer may long for Tahiti, Paris, and Rio de Janiero, so will the student of the origin of life dream of places such as Titan, Europa, and Mars. Neither of the above lists, of course, is meant to be complete. The worlds I have named are a sample, offering a diversity of environments with the potential for chemical evolution. No firm plans have been made, at the time of this writing, for the further exploration of any of them by humans, or even by robot landers. As a substitute, we will visit them in our imaginations.

Titan

At noon on Titan, the largest moon of Saturn, the distant sun shines dimly through a red-orange smog, affording only as much light as a full moon does on earth. The eerie glow reveals an enormous sea, whose gentle waves wash the shores of a continent. At times, storms occur. Rains fall on the land, feeding rivers that cut a path through the soil and flow to the sea. The rich, thick atmosphere is largely made of nitrogen gas.

The details of the landscape given thus far may remind us of earth, but the similarities end at this point. The atmosphere, denser than that of earth, contains in addition to nitrogen, some argon, a few percent of methane, and a fraction of a percent of hydrogen. Its reducing character resembles that of early models of the primitive earth.

In that atmosphere, light and electrical discharges interact with the various gases to afford a gigantic Miller-Urey experiment. A number of the molecules present in interstellar dust clouds are produced, including hydrogen cyanide, hydrocarbons, and nitrogenous organic compounds. Further combination of these substances produces organic particles, which drift slowly out of the atmosphere. These particles accumulate on land, creating a meters-thick layer of soil, perhaps better described as soot.

Above all, a frightful chill pervades the place. The temperature everywhere on Titan hovers near -178°C (-288°F), a value closer to the near absolute cold of outer space than to the worst Siberian winter on earth. All of the water on Titan has frozen to ice, which constitutes the bedrock of the continents. The clouds, rain, rivers, and sea are made of methane and other hydrocarbons.

Organic molecules, formed in the air, are free to enter the hydrocarbon sea. They can interact with one another, and even undergo a degree of chemical evolution. The familiar reactions of earth would go extremely slowly, due to the very low temperature. However, other substances, too fragile to survive the heat of earth, might serve for evolution in the cold hydrocarbon sea.

Our world, of course, provides hot spots, vents, and volcano craters with temperatures well above the average of the surface. Titan may do so as well. The equivalent on Titan of a lava flow would be a stream of water emerging from the warmer interior. For short times, in select locations, liquid water may interact with organic molecules to produce reactions of a type more common to the planet Earth.

Much of the above account is speculation, taken from conjectures in technical articles published by others, or supplied by myself. Titan is larger than some planets. Its distance and its thick cloud cover are obstacles to direct observation from earth. Much of our information comes from the single fly-by of Voyager I in November 1980.

Estimates of Titan's temperature and of the general nature of its atmosphere appear fairly secure. The hydrocarbon sea and rains have been the subject of debate and may or may not exist. The molecules described have been detected. Their further reaction products may be of no more interest to the origin of life than the asphalt used to pave some of our roads. Alternatively, a principle of gradual chemical evolution may have taken over and produced an evolving, organized system of the type important for us to study. Some in the origin-of-life field feel that amino acids and nucleic acid components, or the polymers themselves, would be the expected products of such a process. I myself feel that Titan is not like the earth, and that if chemical evolution has occurred it would most likely have taken some other path.

Fortunately, this dispute concerns science, and not religion or mythology. We have within our power the means to learn what we wish about this world, initially by remote observation and ultimately by a direct visit. We need not wait until judgment day for the answer.

Europa

The four largest moons of Jupiter—Io, Europa, Ganymede, and Callisto—have been known to mankind since their discovery by Galileo in 1610. Most of our knowledge of them, however, came from fly-bys of NASA spacecraft in the 1970s, particularly the encounter with Voyager I and II in 1979.

Io, the closest to Jupiter, has the appearance of a pizza pie, with active volcanoes, and characteristics unlike the other three. The remainder have surfaces of ice, essentially no atmosphere, and densities that indicate that they are made of both ice and rock. If they had undergone a process of differentiation (internal melting, with the heaviest components settling to the center) during their formation, as earth most likely has done, then the rock would make up their cores, with the ice as a mantle above them.

The temperature of this ice layer is the feature of most interest to us. The surface ice, exposed to space, has a temperature of about -170°C (-274°F). If, however, any of the three moons has an internal heat supply due to radioactivity, as does the earth, then some or all of this ice mantle may instead be in molten form, as water. An internal ocean would exist which could provide a suitable site for chemical evolution, and perhaps for the generation of life based on carbon chemistry and water, as is our own.

In an earlier book, a colleague, physicist Gerald Feinberg, and I discussed this possibility in terms of the largest moon of Jupiter, Ganymede. More recently, attention has shifted to Europa, the second closest to Jupiter of the large satellites.

Europa is slightly smaller than our own moon, and its density indicates that perhaps 6 percent of its mass is water, less than that of Ganymede or Callisto. Europa, however, presents a surface unlike the others, with many filled-in cracks but few impact craters. As Europa has undoubtedly undergone the same bombardment by meteorites as have other bodies in the solar system, the craters presumably have been reabsorbed by some process. These features have been interpreted by scientists from NASA and the University of California, Santa Barbara, as evidence for an internal ocean under a thin, somewhat elastic crust of ice. The ocean could be more than 100 kilometers (60 miles) deep. Tidal forces, due to the interaction of Europa with Jupiter, as well as radioactivity, would produce the heat to keep the water in liquid form.

An energy source would be needed to drive chemical evolution and sustain life. How would a suitable one be found in the dark ocean beneath the icy crust of Europa? The NASA scientists, David Reynolds and Steven Squyres, suggested that transient small cracks in the ice crust would allow sunlight to penetrate the ocean for periods of three or four years, providing energy at intervals. In our book, Gerald Feinberg and I presented an alternative. Hydrothermal vents could line the floor of such an ocean, as they do on earth. If the vents are sufficient to sustain life here, independently of sunlight, and are even the favored site of some scientists for the origin of life on earth, why could they not play the same role on Europa?

There may be life under the ice of Europa. If the internal ocean exists, it may have been there for billions of years, sufficient for substantial evolution to occur. To find out what has happened, we must peek under this crust, an expensive undertaking. In 1989, if the current launch schedule holds, an orbiter of a NASA mission to Jupiter, Project Galileo, will inspect the surface of the various satellites in greater detail. Possibly, more clues concerning the existence of an internal ocean will be found. If organic material in substantial amounts had erupted from the ocean and spilled onto the surface, this might also be detected. A more complete answer concerning the contents of Europa will require a lander, a project for the twenty-first century.

Mars

We already have had a chance to inspect the planet Mars at close range. Two identical landers of the Viking project were placed at widely separated points on the surface in July 1976. They attempted, through various tests, to detect microbial life similar to our own. The results were ambiguous and confusing. Something interesting exists on the surface of Mars, but we do not know what it is. Only an additional lander mission, or missions, will tell us.

Conditions at the Viking lander sites, at first glance, do not seem hospitable to our type of life. The Viking cameras revealed dry, red-orange, boulder-strewn deserts. The temperature varied from -90° to -10°C (-130° to +14°F), remaining below the freezing point of water. Liquid water is not present on the surface there, or anywhere else on Mars, though ice exists in the polar cap, traces of water vapor are present in the air, and some water is bound to minerals in the soil. The Martian atmosphere is about 1 percent as dense as our own and is mainly nitrogen, with some argon and carbon dioxide. These unpleasant conditions, and the barren appearance of Mars, did not prohibit the existence of microbial life there, however. It was the task of a chemical analysis instrument, and three separate biological experiments, to detect it, if it existed.

The biological results were on the whole encouraging. Three different types of chemical changes, typical of metabolism of microorganisms on earth, were sought. Each test involved different assumptions, and it was agreed in advance that a positive response on any of the three would be considered a good indication of the presence of life. One experiment, in fact, gave clearly positive results. Carbon dioxide was released when a solution of simple organic compounds was applied to Martian soil. The other two biological tests gave results that were neither clearly positive nor negative, within the initial conception of the experiments. Oxygen gas was released, for example, when Martian soil was treated with water, a totally unexpected outcome.

Taken alone, the biological tests would have suggested that life was present in the soil samples. The chemical analysis instrument, however, detected the presence of no organic compounds at all. On earth, organisms in the soil are inevitably accompanied by additional organic matter, which is readily detected.

A number of explanations have been put forth to resolve this apparent paradox. The positive biological test is more sensitive to the presence of microorganisms than the chemical instrument. Thus a low level of microbes could have been picked up by the one and missed by the other. Most scientists prefer a more conservative, nonbiological explanation for all of the results, however. Hosts of inorganic chemical systems have been explored, with only partial success, in an attempt to simulate the Viking results.

Amazingly, some of the better results in simulating the Viking life-detection experiments were obtained in systems based on layered clay minerals. Efforts to explain away the possible presence of life on Mars may have taken us into the very systems responsible for the origin of life on earth. Active clay-mineral organisms of the type described by Graham Cairns-Smith would of course fit all of the Viking results, including the organic analysis experiment. It would be most ironic if we had to travel to Mars to meet our most ancient ancestors.

Life, or chemical evolution, elsewhere on Mars would not be precluded, even if the Viking soil samples were actually lifeless, and the results were due to the most boring set of chemical reactions imaginable. The closest earthly analog of a Martian environment exists in certain cold, dry, windswept deserts of Antarctica that were discussed earlier. Bacteria and algae live comfortably there, hidden just below the surface of rocks. Life may also exist at the Viking sites, but within the rocks or at a deeper level in the soil, beyond the reach of the scoop used on the lander. Gilbert Levin, the member of the Viking project who devised the most successful biological experiment, noticed green patches on nearby rocks that resembled lichen, but could not persuade other members of the team to take an interest in them.

Even if the entire vicinity of the landers should prove to be of no interest, other locations might still hold promise. For example, certain sites below the equator of Mars may contain liquid water just beneath the surface. The edge of the polar icecap would be another place of interest. We will want to examine a good portion of the planet, perhaps using a remote-controlled mobile vehicle, a rover, before we can be sure what might or might not be there.

If present life were absent, we might yet encounter relics of past life. The presence of many ancient riverlike channels suggests that Mars, early in its history, had water on its surface. Wind or ice have been suggested as alternative agents for the formation of these channels, but the river explanation seems plausible. Life, then, could have evolved on Mars in a much earlier moist and warm period, and died out as the climate changed. If so, we may encounter fossils which represent this phase of Martian history.

Mars may hold important lessons for us concerning the origin and evolution of life and its extent in the universe. In the Viking project, we attempted to get the information by a stab-in-the-dark, one-shot effort. This attempt failed, but may have taught us an important lesson. A sustained and patient effort, even in the face of adversity, will be needed to learn the full story of Mars. We will probably not be satisfied until human beings have walked the dry riverbeds and dug beneath the surface of the planet. Should results for the origin-of-life question be entirely negative, we would still be very proud of the way in which the search was carried out.

Planetary Adventures

After a period of virtual eclipse in the early 1980s, the planetary exploration program is now showing modest signs of revival. An advisory committee to NASA on solar system exploration recommended a moderate program for the rest of this century. Fourteen core missions were listed which could operate within current budget restrictions. Of these fourteen, four were selected for special emphasis. One of the four had special significance for studies of the origin of life, a Titan Probe–Radar Mapper. A probe would descend by parachute into the atmosphere of Titan, determining its exact composition and mapping a portion of the surface.

A mission of this type, called Project Cassini, has received high priority from the European Space Agency as well. If the two agencies cooperate, a joint reconnaissance of the Saturn system might be made, with the exploration of Titan a key feature; a possible target date would be 1995. Possibly, success in this venture would set the stage for more ambitious space explorations in the early twenty-first century—for example, an extensive study of the surface of Mars, culminating in a manned expedition.

In the short run, such explorations may or may not give us important insights concerning the principles of chemical evolution and the origin of life. The spirit that motivates them, if not suppressed, will ultimately take us, or probes sent by us, out beyond our solar system into the vaster reaches of the galaxy. Out there, our descendants will surely find answers to the questions concerning the place of life in the universe. In the interim, we of this generation must be content with the partial answers that may become available to us.

A Guess

Although the full story of our origins is yet to come, I hesitate to leave a vacuum: We need some model to organize the material in hand, locate discrepancies, and plan further studies. In the following account, I will try to tie together the various threads we have collected and sketch in the blanks. I hope that my ideas are not taken as dogma, as the field has no need of additional mythology.

I would assume, to start, that the life we know is the product of our own planet. We have hardly examined the opportunities here as yet—there is no need to turn elsewhere. The simplest assumption about conditions on earth before life began is that they were much as they are today. The exception, of course, is that life and its products, most notably the oxygen in the air, were absent. Further, no steps of great improbability were involved, just predictable developments, which would occur again under the same circumstances. In other environments, with different circumstances, other chemical paths would be followed and different forms of life would arise, or no life at all.

The complicated molecules and structures that we observe in life today are presumably the result of a long process of evolution, just as the organs of our society, Congress, the courts, the Internal Revenue Service, result from a long period of social development. It makes no more sense to presume that life began with well-developed enzymes and replicating systems than it does to assume that primitive tribes arranged complex legislatures and tax bureaus when they first learned to govern themselves. Life began with simple, available chemicals and then progressed.

The problem that remains is to specify the ingredients, circumstances, and organizing principles. This exercise must be more a matter of intuition than logic, given the present state of our knowledge. I myself was quite taken by an area I observed while on vacation in Yellowstone National Park in 1983. The location had the provocative name "Fountain Paint Pot," and was one of many geothermal sites in that vicinity. I walked past deep-blue hot pools, spouting geysers, and boiling springs. As the waters ran downhill from these features, they deposited bright yellow-orange and red streaks of sulfur, iron oxides, and other minerals on the rocks below. The sun shone brightly on this color display, but the air was less congenial as it carried the unpleasant odor of reduced sulfur gases.

The area drew its name from its central attraction, a frothy pool of dense mud, or mudpot. Steam bubbled through the thick mud, made of kaolinite clay, hurling material into the air and producing ripples on the thick, viscous surface. As I watched this busy, almost sensual display, I thought, This must be the place. It was rare to see inanimate matter act so animate.

Abundant energy was available here, in the form of sun, wind, heat, chemicals, and flowing water. Reduced chemicals were present, and minerals were tossed about in the sunlight. Surely, many sites like this were present on the primitive earth, and at one or more of them, something happened.

Many others have had this idea before me, but no one has been able to deduce the exact steps or the principles that were involved. Clays were present in abundance; we have discussed their possibilities. We are made of organic chemicals, so they must have entered at some point. They came in not as evolved products, but as simple ones, of a few carbon atoms. Otherwise, the chemical complexity would have been overwhelming. It is not yet clear whether clays evolved alone for a time, as Graham Cairns-Smith suggests, or whether the partnership of carbon and silicate functioned from the very beginning.

The intermingling of these two sources of chemical diversity was apparent at Fountain Paint Pot. The pigments were not provided by minerals alone. Green, orange, and brown algae colored the hot water, as did yellow and pink bacteria. My unpracticed eye could not tell the minerals from the microorganisms.

They differ vastly, of course, as we discussed in earlier chapters. Their proximity in this place, however, suggested some earlier collaboration in the origin of life. The organic partner in this early union gained in complexity and evolved, while its mineral companion remained in place.

The Joy of Science

If the above account should turn out to be accurate, I would be quite surprised. I have tried to bring together the most relevant things that we have learned, but I could not include the discoveries not yet made. Science is not the place for those who want certainty, who wish the truths they learned in childhood to reassure them in their old age. Surprises occur, and alter our perception of reality—for example, the discovery of radioactivity or the genetic role of DNA.

Some areas of science, such as classical mechanics and basic organic chemistry, seem reasonably well settled. Fundamental discoveries in such areas are possible, but need not be expected. In the origin of life, however, if no surprises were forthcoming, that would be the most surprising result of all.

Whatever discoveries the future may hold, interest in this topic will endure. Perhaps some human beings are content to live their lives on a day-to-day basis without wondering about the large questions of science. How did the universe begin? How did life start in it? What kinds of life exist? How does consciousness work? Such an attitude would remind me of an amnesia victim who awoke one day with no past memory, and took no interest in his earlier life. Francis Crick has written: "To show no interest in these topics is to be truly uneducated."

But in fact, people *have* been interested in the question of origins since ancient times, and remain so to this day. Lecture halls, otherwise sparsely populated, fill up when the origin of life is the announced topic. Many of those attending hope that the answer will be given on that very day. If they come for that reason alone, then they are fated to be disappointed. But other valid reasons exist for their attention, if they come because of an interest in science. If so, they come not only to satisfy their curiosity about the answer, but also because they enjoy the spirit of the quest.

When we treat each new observation and theory with skepticism, retaining our doubt until it has passed the test of experience, and then place it alongside our other acquisitions with the care of a collector who has acquired a valued object after a long search, then we can experience the joy of science. It is this joy, rather than an insistence on an immediate answer, that is likely to be our reward as we continue to search for the origin of life. But even in this conclusion, let us exercise some caution. We may be closer to the answer than we think.

Further Reading

A number of references that I found useful are compiled here, by chapter. The reader who wishes to learn more about topics discussed in the text should consult them. This list is intended to be suggestive rather than comprehensive.

Chapter 1

Sun Songs. Creation Myths from Around the World, Raymond Van Over, Ed. (New York: New American Library, 1980), provides a valuable introduction to this subject.

The career of Ignaz Semmelweis is described in *Immortal Magyar*, by Frank G. Slaughter (New York: Henry Schulman, 1950).

For information on the philosophy and practice of science, see Thomas Kuhn, *The Structure of Scientific Revolutions*, 2d ed. (Chicago: University of Chicago Press, 1970), and Carl G. Hempel, *Philosophy of Natural Science* (Englewood Cliffs, N.J.: Prentice-Hall, 1966).

The Spontaneous Generation Controversy from Descartes to Oparin, by John Farley (Baltimore: Johns Hopkins University Press, 1977), provides an excellent history of the subject.

CHAPTER 2

Additional discussion on the nature of life can be found in G. Feinberg and R. Shapiro, *Life Beyond Earth* (New York, William Morrow, 1980). See also the article "Life" by Carl Sagan in the *Encyclopaedia Britannica*, 15th ed., *Macropaedia*, Vol. 10.

CHAPTER 3

The basis of the calculation by Archbishop Ussher is described in detail by William R. Brice in "Bishop Ussher, John Lightfoot and the Age of Creation," *Journal of Geological Education*, 30 (1982), 18–24.

An account of various methods used to estimate the age of the earth, including radioactive dating, is given in Frank Press and Raymond Siever, *Earth*, 3d ed. (San Francisco: W. H. Freeman, 1982).

Information on the controversies surrounding the Isua "fossils" can be found in articles by D. Bridgwater, J. H. Aalaart, J. W. Schopf, C. Klein, M. R. Walter, E. S. Barghoorn, P. Strother, A. H. Knoll and B. E. Gorman, "Microfossil-like Objects from the Archaean of Greenland: a Cautionary Note," *Nature*, 289 (1981) 51–52, and B. Nagy, M. H. Engel, John M. Zumberge, H. Ogino, and S. Y. Chang, "Amino Acids and Hydrocarbons 3,8000-Myr Old in the Isua Rocks, Southwestern Greenland," ibid, 53–56, as well as the reference cited therein.

For comprehensive information concerning the geological history of the earth and its relation to the development of life, see *Earth's Earliest Biosphere*, J. William Schopf, Ed. (Princeton, N.J.: Princeton University Press, 1983).

CHAPTER 4

The circumstances surrounding his early experiment are described in an account by Stanley Miller, "The First Laboratory Synthesis of Organic Compounds under Primitive Earth Conditions," in *The Heritage of Copernicus: Theories "Pleasing to the Mind,"* J. Neyman, Ed. (Cambridge, Mass.: MIT Press, 1974), pp. 228–242.

A summary of Miller's results can be found in Stanley L. Miller and Leslie E. Orgel, *The Origins of Life on the Earth* (Englewood Cliffs, N.J.: Prentice-Hall, 1974). See also the references on pp. 100–101 of that work. His more recent work with less favorable atmospheres is described by S. L. Miller and G. Schlesinger in "Carbon and Energy Yields in Prebiotic Syntheses using Atmospheres Containing CH_4, CO, and CO_2," *Origins of Life*, 14 (1984) 83–89. For a critical analysis of the significance of these experiments, see A. G. Cairns-Smith, *Ge-*

netic Takeover and the Mineral Origins of Life (New York: Cambridge University Press, 1982).

The relation between the amino acids found in meteorites and those produced in Miller-Urey experiments is described by James G. Lawless and Etta Peterson in "Amino Acids in Carbonaceous Chondrites," *Origins of Life*, 6 (1976), 3–8.

A summation of the ideas of A. I. Oparin is presented by him in *Life, Its Nature, Origin and Development*, translated from the Russian by A. Synge (New York: Academic Press, 1964).

Criticisms of the "primeval [prebiotic] soup" concept are presented by J. Brooks and G. Shaw in "A Critical Assessment of the Origin of Life," published in *Origins of Life*, 9 (1978), pp. 597–606.

Carl Woese's proposal of a cloud location for the origin of life, "An Alternative to the Oparin View of the Primeval Sequence," was published in *The Origins of Life and Evolution*, H. O. Halvorson and K. E. Van Holde, Eds. (New York: Alan R. Liss, 1980), pp. 65–76.

The bacteria said to grow at 250° C are discussed by J. D. Trent, R. A. Chastain and A. A. Yayanos in "Possible Artefactual Basis for Apparent Bacterial Growth at 250° C," *Nature*, 307 (1984), pp. 737–740. See also the reply by J. A. Baross and J. W. Deming on p. 740.

Chapter 5

The August 1954 *Scientific American* article by George Wald, "The Origin of Life," is reprinted in *Life, Origin and Evolution*, with introduction by Clair E. Folsome (San Francisco: W. H. Freeman, 1979).

The calculations by Harold Morowitz are discussed in detail in his book, *Energy Flow in Biology* (New York: Academic Press, 1968).

Chapter 6

The article by H. J. Muller cited in the text, "The Gene Material as the Initiator and the Organizing Basis of Life," was published in the *American Naturalist*, 100 (1966), 493–517. The cited article by Carl Sagan, "Radiation and the Origin of the Gene," was published in *Evolution*, 11 (1957), 40–55.

Genes, Radiation and Society. The Life and Work of H. J. Muller, by Elof Carlson (Ithaca, N.Y.: Cornell University Press, 1981), provides an engrossing account of Muller's adventures.

For information about the career of Lysenko and its effect on Soviet science, see David Joravsky, *The Lysenko Affair* (Cambridge, Mass.: Harvard University Press, 1970), Loren A. Graham, *Science and Philosophy in the Soviet Union* (New York: Knopf, 1972), and Zhores A. Medvedev, *The Rise and Fall of T. D. Lysenko,* translated by I. Michael Lerner (New York: Columbia University Press, 1969).

A work by A. I. Oparin that summarizes his ideas has been listed in the references for Chapter 4.

Proceedings of the First International Symposium on "The Origin of Life on the Earth," F. Clark and R. L. M. Synge, Eds. (New York: Pergamon Press, 1959).

The work on spontaneous generation by Farley is listed under Chapter 2.

Chapter 7

The Darwinian experiments conducted by Sol Spiegelman and his co-workers are described in F. R. Kramer, D. R. Mills, P. E. Coles, T. Nishihara and S. Spiegelman, "Evolution in vitro: Sequence and Phenotype of a Mutant RNA Resistant to Ethidium Bromide," in *Journal of Molecular Biology,* 89 (1974) 719–736, and in references cited therein.

For information on the work of Leslie Orgel on models of RNA replication, see T. Inoue and L. E. Orgel, "A Nonenzymatic RNA Polymerase Model" in *Science,* 219 (1983), 859–862.

Manfred Eigen and his collaborators have summarized their ideas on the origin of life in "The Origin of Genetic Information," by M. Eigen, W. Gardiner, P. Schuster, and R. Winkler-Oswatitsch, *Scientific American,* April 1981.

The theory of Norman Horowitz that is discussed was published as an article, "On the Evolution of Biochemical Synthesis," in *Proceedings of the National Academy of Sciences,* 31 (1945), 152–157.

The constraints followed in prebiotic syntheses have been summarized by L. E. Orgel and R. Lohrmann in "Prebiotic Chemistry and Nucleic Acid Replication," *Accounts of Chemical Research,* 7 (1974), 368–377.

The literature on prebiotic synthesis is vast. The following references represent a selection of those used to construct the revised version of Wednesday's Tale: A. W. Schwartz, "Chemical Evolution—the Genesis of the First Organic Compounds," *Marine Organic Chemistry,* E. K. Duursma and R. Dawson, Eds. (Amsterdam: Elsevier, 1981), pp. 7–30; N. W. Gabel and C. Ponnamperuma, "Model for Origin of Monosaccharides," *Nature,* 216 (1967), 453–455; W. D. Fuller, R. A. Sanchez and L. E. Orgel, "Studies in Prebiotic Synthesis. VII. Solid State Synthesis of Purine Nucleosides," *Journal of Molecular Evolution,* 1 (1972) 249–257; M. J. Bishop, R. Lohrmann and L. E. Orgel, "Prebiotic Phosphorylation of Thymidine at 65° C in Simulated Desert

Conditions," *Nature,* 237 (1972) 162–164; D. A. Usher, "Early Chemical Evolution of Nucleic Acids: A Theoretical Model," *Science,* 196 (1977) 311–313.

CHAPTER 8

A comprehensive discussion of the work of Sidney Fox and his collaborators is included in S. Fox and K. Dose, *Molecular Evolution and the Origin of Life,* rev. ed. (New York: Marcel Dekker, 1977). For a more recent summary, see S. Fox in *Science and Creationism,* Ashley Montague, Ed. (New York: Oxford University Press, 1984), pp. 194–239. A critical point of view is presented by William Day in *Genesis on Planet Earth* (East Lansing, Mich.: House of Talos, 1979).

Oscillations in chemical reactions are described by I. R. Epstein, K. Kustin, P. De Kepper, and M. Orban in "Oscillating Chemical Reactions," *Scientific American,* March 1983.

The ideas of A. G. Cairns-Smith are fully described in his recent book: *Genetic Takeover and the Mineral Origins of Life* (New York: Cambridge University Press, 1982).

CHAPTER 9

Francis Crick's theory of Directed Panspermia is described in detail in *Life Itself* (New York: Simon and Schuster, 1981).

Popular books by F. Hoyle and N. C. Wickramasinghe include *Lifecloud* and *Diseases from Space* (New York: Harper & Row, 1978, 1979) and *Evolution from Space* (New York: Simon and Schuster, 1981). More recently Hoyle has written *The Intelligent Universe* (New York: Holt, Rinehart and Winston, 1984).

For a sample of the technical papers of H. and W. on interstellar grains, see F. Hoyle, A. H. Olavesen and N. C. Wickramasinghe, "Identification of Interstellar Polysaccharides and Related Hydrocarbons," *Nature,* 271 (1978) 229–231, and F. Hoyle and N. C. Wickramasinghe, "Biochemical Chromophores and the Interstellar Extinction at Ultraviolet Wavelengths," *Astrophysics and Space Science,* 65 (1979) 241–244.

CHAPTER 10

For a brief history of the Creationist movement in the United States, see Ronald L. Numbers, "Creationism in 20th-Century America," *Science,* 218 (1982) 538–544.

The text of Judge William R. Overton's decision "Creationism in Schools: The Decision in McLean versus the Arkansas Board of Education" has been published in *Science,* 215 (1982) 934–943. It contains historical material, a fascinating account of the trial, and an excellent discussion of the nature of science.

A general account of the Creationist position can be found in *Scientific Creationism*, public school edition (San Diego, Calif., CLP Publishers, 1974). For a more technical discussion of the Creationist point of view on the origin of life, see Duane T. Gish, "A Consistent Christian-Scientific View of the Origin of Life," *Creation Research Society Quarterly*, 15 (1979) 185–203.

Arguments against the Creationist view of science may be found in Niles Eldridge, *The Monkey Business* (New York: Washington Square Press, 1982), and Philip Kitcher, *Abusing Science* (Cambridge, Mass.: MIT Press, 1982).

CHAPTER 11

The proceedings of the Seventh International Conference on the Origin of Life have been published in *Origins of Life*, vol. 14, no. 1–4 (1984).

CHAPTER 12

For a recent review of advances in DNA sequencing, see A. T. Bankier, "Advances in Dideoxy Sequencing," *BioTechniques*, March/April 1984, pp. 72–77, and references therein. The largest entity sequenced to early 1985 is the Epstein-Barr virus genome of 172,282 base pairs; see R. Baer, A. T. Bankier, M. D. Biggin, P. L. Deininger, P. J. Farrell, T. J. Gibson, G. Hatfull, G. S. Hudson, S. C. Satchwell, C. Seguin, P. S. Tuffnell and B. G. Barrell, "DNA Sequence and Expression of the B95-8 Epstein-Barr Virus Genome," *Nature*, 310 (1984) 207–211.

Information on scrapie may be found in an article by Stanley B. Prusiner, "Prions," *Scientific American*, October 1984.

CHAPTER 13

An extensive criticism of the design of many prebiotic simulations has been made by C. B. Thaxton, W. L. Bradley and R. Olsen, *The Mystery of Life's Origin: Reassessing Current Theories* (New York: Philosophical Library, 1984).

The possibility of life elsewhere in our solar system has been discussed at length by a colleague and myself: G. Feinberg and R. Shapiro, *Life Beyond Earth* (New York: William Morrow, 1980).

The Planetary Report, vol. III, no. 6 (1983), is almost entirely devoted to Titan, including plans for future exploration.

The possibility of liquid water within Europa is considered by S. W. Squyres, R. T. Reynolds and P. M. Cassen in "Liquid Water and Active Resurfacing on Europa," *Nature*, 301 (1983) 225–226.

In the cover story of its September 1984 issue, *Discover* magazine featured the arguments for a manned voyage to Mars.

Index

About the Author

Robert Shapiro is a Professor of Chemistry at New York University and an expert on DNA research and on the genetic effect of environmental chemicals. He is the co-author with Gerald Feinberg of *Life Beyond Earth,* which *The New York Times Book Review* called "one of the best books on Earth about life elsewhere."